Wildlife Habitat Management

Wildlife Habitat Management

Edited by
Vincent Green

Larsen & Keller
www.larsen-keller.com

Wildlife Habitat Management
Edited by Vincent Green
ISBN: 978-1-63549-698-7 (Hardback)

Larsen & Keller

Published by Larsen and Keller Education,
5 Penn Plaza,
19th Floor,
New York, NY 10001, USA

Cataloging-in-Publication Data

Wildlife habitat management / edited by Vincent Green.
      p. cm.
Includes bibliographical references and index.
ISBN 978-1-63549-698-7
1. Wildlife habitat improvement. 2. Wildlife management.
I. Green, Vincent.
QL82 .W55 2018
639.92--dc23

For more information regarding Larsen and Keller Education and its products, please visit the publisher's website www.larsen-keller.com

Table of Contents

Preface

Earth's biodiversity is at risk because of deforestation, population increase and increasing pollution. Wildlife habitat management is the discipline which focuses on preserving the wildlife from the ill effects of human actions and creating and maintaining a balance between the wildlife and human civilization. This subject uses elements of geography, ecology, chemistry, climatology, biology, etc. This book unfolds the innovative aspects of wildlife habitat management which will be crucial for the holistic understanding of the subject matter. Most of the topics introduced in it cover new techniques and the applications of the area. This textbook, with its detailed analyses and data, will prove immensely beneficial to professionals and students involved in this area at various levels.

A detailed account of the significant topics covered in this book is provided below:

Chapter 1- The areas in which an organism lives is known as its habitat. Polar, subtropical, temperate are some types of habitats. With time, habitats witness change. These changes can be brought by events such as earthquakes, wildfires, volcano eruptions, etc. This chapter will provide an integrated understanding of habitats.

Chapter 2- There are various types of habitats that contain different species of flora and fauna. Habitats are unique and suited to the organisms that make them their home. Some types of habitats include marine habitats, deserts, etc. The topics discussed in the chapter are of great importance to broaden the existing knowledge on habitats.

Chapter 3- Extinction is the termination of a group of organisms or a species so that they cease to exist. Extinctions can be of two types, mass extinctions and isolated extinctions. Climate change and human intervention that results in habitat destruction are some of the main causes of extinction. The chapter on ecological extinction offers an insightful focus, keeping in mind the complex subject matter.

Chapter 4- Animals which are not domesticated are a part of wildlife. A broader definition would be organisms that are found in the wild. It includes various types of animals, plants and fungi. Wildlife trade, wildlife conservation, conservation biology, habitat conservation and conservation movement are some of the topics related to wildlife conservation. The aspects elucidated in this chapter are of vital importance, and provide a better understanding of wildlife habitat management.

Chapter 5- Marine conservation is the conservation of seas and oceans and the organisms found in them. The aim of this practice is to restrict the damage caused by humans to marine ecosystems. The topics discussed in the chapter are of great importance to broaden the existing knowledge on marine conservation.

It gives me an immense pleasure to thank our entire team for their efforts. Finally in the end, I would like to thank my family and colleagues who have been a great source of inspiration and support.

Editor

An Introduction to Habitat

The areas in which an organism lives is known as its habitat. Polar, subtropical, temperate are some types of habitats. With time, habitats witness change. These changes can be brought by events such as earthquakes, wildfires, volcano eruptions, etc. This chapter will provide an integrated understanding of habitats.

Habitat

This coral reef in the Phoenix Islands Protected Area is a rich habitat for sea life.

A habitat is an ecological or environmental area that is inhabited by a particular species of animal, plant, or other type of organism. The term typically refers to the zone in which the organism lives and where it can find food, shelter, protection and mates for reproduction. It is the natural environment in which an organism lives, or the physical environment that surrounds a species population.

A habitat is made up of physical factors such as soil, moisture, range of temperature, and light intensity as well as biotic factors such as the availability of food and the presence or absence of predators. Every organism has certain habitat needs for the conditions in which it will thrive, but some are tolerant of wide variations while others are very specific in their requirements. A habitat is not necessarily a geographical area, it can be the interior of a stem, a rotten log, a rock or a clump of moss, and for a parasitic organism it is the body of its host, part of the host's body such as the digestive tract, or a single cell within the host's body.

Habitat types include polar, temperate, subtropical and tropical. The terrestrial vegetation type may be forest, steppe, grassland, semi-arid or desert. Fresh water habitats include marshes, streams, rivers, lakes, ponds and estuaries, and marine habitats include salt marshes, the coast, the intertidal zone, reefs, bays, the open sea, the sea bed, deep water and submarine vents.

Few creatures make the ice shelves of Antarctica their habitat.

Habitats change over time. This may be due to a violent event such as the eruption of a volcano, an earthquake, a tsunami, a wildfire or a change in oceanic currents; or the change may be more gradual over millennia with alterations in the climate, as ice sheets and glaciers advance and retreat, and as different weather patterns bring changes of precipitation and solar radiation. Other changes come as a direct result of human activities; deforestation, the ploughing of ancient grasslands, the diversion and damming of rivers, the draining of marshland and the dredging of the seabed. The introduction of alien species can have a devastating effect on native wildlife, through increased predation, through competition for resources or through the introduction of pests and diseases to which the native species have no immunity.

Ibex in alpine habitat

Definition and Etymology

The word "habitat" has been in use since about 1755 and derives from the Latin third-person singular present indicative of *habitāre*, to inhabit, from *habēre*, to have or to hold. Habitat can be defined as the natural environment of an organism, the place in which it is natural for it to live and grow. It is similar in meaning to a biotope; an area of uniform environmental conditions associated with a particular community of plants and animals.

Environmental Factors

The chief environmental factors affecting the distribution of living organisms are temperature, humidity, climate, soil type and light intensity, and the presence or absence of all the requirements that the organism needs to sustain it. Generally speaking, animal communities are reliant on specific types of plant communities.

Some plants and animals are generalists, and their habitat requirements are met in a wide range of locations. The small white butterfly (*Pieris rapae*) for example is found on all the continents of the world apart from Antarctica. Its larvae feed on a wide range of *Brassicas* and various other plant species, and it thrives in any open location with diverse plant associations. The large blue butterfly is much more specific in its requirements; it is found only in chalk grassland areas, its larvae feed on *Thymus* species and because of complex lifecycle requirements it inhabits only areas in which *Myrmica* ants live.

Disturbance is important in the creation of biodiverse habitats. In the absence of disturbance, a climax vegetation cover develops that prevents the establishment of other species. Wildflower meadows are sometimes created by conservationists but most of the flowering plants used are either annuals or biennials and disappear after a few years in the absence of patches of bare ground on which their seedlings can grow. Lightning strikes and toppled trees in tropical forests allow species richness to be maintained as pioneering species move in to fill the gaps created. Similarly coastal habitats can become dominated by kelp until the seabed is disturbed by a storm and the algae swept away, or shifting sediment exposes new areas for colonisation. Another cause of disturbance is when an area may be overwhelmed by an invasive introduced species which is not kept under control by natural enemies in its new habitat.

Types of Habitat

Rich rainforest habitat in Dominica

Terrestrial habitat types include forests, grasslands, wetlands and deserts. Within these broad biomes are more specific habitats with varying climate types, temperature regimes, soils, altitudes and vegetation types. Many of these habitats grade into each other and each one has its own typical communities of plants and animals. A habitat may suit a particular species well, but its presence or absence at any particular location depends to some extent on chance, on its dispersal abilities and its efficiency as a coloniser.

Wetland Habitats in Borneo

Freshwater habitats include rivers, streams, lakes, ponds, marshes and bogs. Although some organisms are found across most of these habitats, the majority have more specific requirements. The water velocity, its temperature and oxygen saturation are important factors, but in river systems, there are fast and slow sections, pools, bayous and backwaters which provide a range of habitats. Similarly, aquatic plants can be floating, semi-submerged, submerged or grow in permanently or temporarily saturated soils besides bodies of water. Marginal plants provide important habitat for both invertebrates and vertebrates, and submerged plants provide oxygenation of the water, absorb nutrients and play a part in the reduction of pollution.

Marine habitats include brackish water, estuaries, bays, the open sea, the intertidal zone, the sea bed, reefs and deep water zones. Further variations include rock pools, sand banks, mudflats, brackish lagoons, sandy and pebbly beaches, and seagrass beds, all supporting their own flora and fauna. The benthic zone or seabed provides a home for both static organisms, anchored to the substrate, and for a large range of organisms crawling on or burrowing into the surface. Some creatures float among the waves on the surface of the water, or raft on floating debris, others swim at a range of depths, including organisms in the demersal zone close to the seabed, and myriads of organisms drift with the currents and form the plankton.

Desert scene in Egypt

A desert is not the kind of habitat that favours the presence of amphibians, with their requirement for water to keep their skins moist and for the development of their young. Nevertheless, some frogs live in deserts, creating moist habitats underground and hibernating while conditions are adverse. Couch's spadefoot toad (*Scaphiopus couchii*) emerges from its burrow when a downpour occurs and lays its eggs in the transient pools that form; the tadpoles develop with great rapidity, sometimes in as little as nine days, undergo metamorphosis, and feed voraciously before digging a burrow of their own.

Other organisms cope with the drying up of their aqueous habitat in other ways. Vernal pools are ephemeral ponds that form in the rainy season and dry up afterwards. They have their specially-adapted characteristic flora, mainly consisting of annuals, the seeds of which survive the drought, but also some uniquely adapted perennials. Animals adapted to these extreme habitats also exist; fairy shrimps can lay "winter eggs" which are resistant to desiccation, sometimes being blown about with the dust, ending up in new depressions in the ground. These can survive in a dormant state for as long as fifteen years. Some killifish behave in a similar way; their eggs hatch and the juvenile fish grow with great rapidity when the conditions are right, but the whole population of fish may end up as eggs in diapause in the dried up mud that was once a pond.

Many animals and plants have taken up residence in urban environments. They tend to be adaptable generalists and use the town's features to make their homes. Rats and mice have followed man around the globe, pigeons, peregrines, sparrows, swallows and house martins use the buildings for nesting, bats use roof space for roosting, foxes visit the garbage bins and squirrels, coyotes, raccoons and skunks roam the streets. About 2,000 coyotes are thought to live in and around Chicago. A survey of dwelling houses in northern European cities in the twentieth century found about 175 species of invertebrate inside them, including 53 species of beetle, 21 flies, 13 butterflies and moths, 13 mites, 9 lice, 7 bees, 5 wasps, 5 cockroaches, 5 spiders, 4 ants and a number of other groups. In warmer climates, termites are serious pests in the urban habitat; 183 species are known to affect buildings and 83 species cause serious structural damage.

Microhabitats

A microhabitat is the small-scale physical requirements of a particular organism or population. Every habitat includes large numbers of microhabitats with subtly different exposure to light, humidity, temperature, air movement, and other factors. The lichens that grow on the north face of a boulder are different to those that grow on the south face, from those on the level top and those that grow on the ground nearby; the lichens growing in the grooves and on the raised surfaces are different from those growing on the veins of quartz. Lurking among these miniature "forests" are the microfauna, each species of invertebrate with its own specific habitat requirements.

There are numerous different microhabitats in a wood; coniferous forest, broad-leafed forest, open woodland, scattered trees, woodland verges, clearings and glades; tree trunk, branch, twig, bud, leaf, flower and fruit; rough bark, smooth bark, damaged bark, rotten wood, hollow, groove and hole; canopy, shrub layer, plant layer, leaf litter and soil; buttress root, stump, fallen log, stem base, grass tussock, fungus, fern and moss. The greater the structural diversity in the wood, the greater the number of microhabitats that will be present. A range of tree species with individual specimens of varying sizes and ages, and a range of features such as streams, level areas, slopes, tracks, clearings and felled areas will provide suitable conditions for an enormous number of biodiverse plants and animals. For example, in Britain it has been estimated that various types of rotting wood are home to over 1700 species of invertebrate.

For a parasitic organism, its habitat is the particular part of the outside or inside of its host on or in which it is adapted to live. The life cycle of some parasites involves several different host species, as well as free-living life stages, sometimes providing vastly different microhabitats. One such organism is the trematode (flatworm) *Microphallus turgidus*, present in brackish water marshes in the southeastern United States. Its first intermediate host is a snail and the second, a glass shrimp. The final host is the waterfowl or mammal that consumes the shrimp.

Extreme Habitats

An Antarctic rock split apart to show an endolithic lifeform showing as a green layer a few millimetres thick.

Although the vast majority of life on Earth lives in mesophyllic (moderate) environments, a few organisms, most of them microbes, have managed to colonise extreme environments that are unsuitable for most higher life forms. There are bacteria, for example, living in Lake Whillans, half a mile below the ice of Antarctica; in the absence of sunlight, they must rely on organic material from elsewhere, perhaps decaying matter from glacier melt water or minerals from the underlying rock. Other bacteria can be found in abundance in the Mariana Trench, the deepest place in the ocean and on Earth; marine snow drifts down from the surface layers of the sea and accumulates in this undersea valley, providing nourishment for an extensive community of bacteria.

Other microbes live in habitats lacking in oxygen, and are dependent on chemical reactions other than photosynthesis. Boreholes drilled 300 m (1,000 ft) into the rocky seabed have found microbial communities apparently based on the products of reactions between water and the constituents of rocks. These communities have been little studied, but may be an important part of the global carbon cycle. Rock in mines two miles deep also harbour microbes; these live on minute traces of hydrogen produced in slow oxidizing reactions inside the rock. These metabolic reactions allow life to exist in places with no oxygen or light, an environment that had previously been thought to be devoid of life.

The intertidal zone and the photic zone in the oceans are relatively familiar habitats. However the vast bulk of the ocean is unhospitable to air-breathing humans, with scuba divers limited to the upper 50 m (160 ft) or so. The lower limit for photosynthesis is 100 to 200 m (330 to 660 ft) and below that depth the prevailing conditions include total darkness, high pressure, little oxygen (in some places), scarce food resources and extreme cold. This habitat is very challenging to research, and as well as being little studied, it is vast, with 79% of the Earth's biosphere being at depths greater than 1,000 m (3,300 ft). With no plant life, the animals in this zone are either detritivores, reliant on food drifting down from surface layers, or they are predators, feeding on each other. Some organisms are pelagic, swimming or drifting in mid-ocean, while others are benthic, living on or near the seabed. Their growth rates and metabolisms tend to be slow, their eyes may be very large to detect what little illumination there is, or they may be blind and rely on other sensory inputs. A number of deep sea creatures are bioluminescent; this serves a variety of functions including predation, protection and social recognition. In general, the bodies of animals living at great depths are adapted to high pressure environments by having pressure-resistant biomolecules and small organic molecules present in their cells known as piezolytes, which give the proteins the flexibility they need. There are also unsaturated fats in their membranes which prevent them from solidifying at low temperatures.

Dense mass of white crabs at a hydrothermal vent, with stalked barnacles on right.

Hydrothermal vents were first discovered in the ocean depths in 1977. They result from seawater becoming heated after seeping through cracks to places where hot magma is close to the seabed. The under-water hot springs may gush forth at temperatures of over 340 °C (640 °F) and support unique communities of organisms in their immediate vicinity. The basis for this teeming life is chemosynthesis, a process by which microbes convert such substances as hydrogen sulfide or ammonia into organic molecules. These bacteria and Archaea are the primary producers in these ecosystems and support a diverse array of life. About 350 species of organism, dominated by molluscs, polychaete worms and crustaceans, had been discovered around hydrothermal vents by the end of the twentieth century, most of them being new to science and endemic to these habitats.

Besides providing locomotion opportunities for winged animals and a conduit for the dispersal of pollen grains, spores and seeds, the atmosphere can be considered to be a habitat in its own right. There are metabolically active microbes present that actively reproduce and spend their whole existence airborne, with hundreds of thousands of individual organisms estimated to be present in a cubic metre of air. The airborne microbial community may be as diverse as that found in soil or other terrestrial environments, however these organisms are not evenly distributed, their densities varying spatially with altitude and environmental conditions. Aerobiology has been little studied, but there is evidence of nitrogen fixation in clouds, and less clear evidence of carbon cycling, both facilitated by microbial activity.

There are other examples of extreme habitats where specially adapted lifeforms exist; tar pits teeming with microbial life; naturally occurring crude oil pools inhabited by the larvae of the petroleum fly; hot springs where the temperature may be as high as 71 °C (160 °F) and cyanobacteria create microbial mats; cold seeps where the methane and hydrogen sulfide issue from the ocean floor and support microbes and higher animals such as mussels which form symbiotic associations with these anaerobic organisms; salt pans harbour salt-tolerant microorganisms and also *Wallemia ichthyophaga*, a basidomycotous fungus; ice sheets in Antarctica which support fungi *Thelebolus* spp., and snowfields on which algae grow.

Habitat Change

Twenty five years after the devastating eruption at Mount St. Helens, United States, pioneer species have moved in.

Whether from natural processes or the activities of man, landscapes and their associated habitats change over time. There are the slow geomorphological changes associated with the geologic processes that cause tectonic uplift and subsidence, and the more rapid changes associated with earthquakes, landslides, storms, flooding, wildfires, coastal erosion, deforestation and changes in land use. Then there are the changes in habitats brought on by alterations in farming practices, tourism, pollution, fragmentation and climate change.

Loss of habitat is the single greatest threat to any species. If an island on which an endemic organism lives becomes uninhabitable for some reason, the species will become extinct. Any type of habitat surrounded by a different habitat is in a similar situation to an island. If a forest is divided into parts by logging, with strips of cleared land separating woodland blocks, and the distances between the remaining fragments exceeds the distance an individual animal is able to travel, that species becomes especially vulnerable. Small populations generally lack genetic diversity and may be threatened by increased predation, increased competition, disease and unexpected catastrophe. At the edge of each forest fragment, increased light encourages secondary growth of fast-growing species and old growth trees are more vulnerable to logging as access is improved. The birds that nest in their crevices, the epiphytes that hang from their branches and the invertebrates in the leaf litter are all adversely affected and biodiversity is reduced. Habitat fragmentation can be ameliorated to some extent by the provision of wildlife corridors connecting the fragments. These can be a river, ditch, strip of trees, hedgerow or even an underpass to a highway. Without the corridors, seeds cannot disperse and animals, especially small ones, cannot travel through the hostile territory, putting populations at greater risk of local extinction.

Habitat disturbance can have long-lasting effects on the environment. *Bromus tectorum* is a vigorous grass from Europe which has been introduced to the United States where it has become invasive. It is highly adapted to fire, producing large amounts of flammable detritus and increasing the frequency and intensity of wildfires. In areas where it has become established, it has altered the local fire regime to such an extant that native plants cannot survive the frequent fires, allowing it to become even more dominant. A marine example is when sea urchin populations "explode" in coastal waters and destroy all the macroalgae present. What was previously a kelp forest becomes an urchin barren that may last for years and this can have a profound effect on the food chain.

Removal of the sea urchins, by disease for example, can result in the seaweed returning, with an over-abundance of fast-growing kelp.

Habitat Protection

The protection of habitats is a necessary step in the maintenance of biodiversity because if habitat destruction occurs, the animals and plants reliant on that habitat suffer. Many countries have enacted legislation to protect their wildlife. This may take the form of the setting up of national parks, forest reserves and wildlife reserves, or it may restrict the activities of humans with the objective of benefiting wildlife. The laws may be designed to protect a particular species or group of species, or the legislation may prohibit such activities as the collecting of bird eggs, the hunting of animals or the removal of plants. A general law on the protection of habitats may be more difficult to implement than a site specific requirement. A concept introduced in the United States in 1973 involves protecting the critical habitat of endangered species, and a similar concept has been incorporated into some Australian legislation.

International treaties may be necessary for such objectives as the setting up of marine reserves. Another international agreement, the Convention on the Conservation of Migratory Species of Wild Animals, protects animals that migrate across the globe and need protection in more than one country. However, the protection of habitats needs to take into account the needs of the local residents for food, fuel and other resources. Even where legislation protects the environment, a lack of enforcement often prevents effective protection. Faced with food shortage, a farmer is likely to plough up a level patch of ground despite it being the last suitable habitat for an endangered species such as the San Quintin kangaroo rat, and even kill the animal as a pest. In this regard, it is desirable to educate the community on the uniqueness of their flora and fauna and the benefits of ecotourism.

Monotypic Habitat

A monotypic habitat is one in which a single species of animal or plant is so dominant as to virtually exclude all other species. An example would be sugarcane; this is planted, burnt and harvested, with herbicides killing weeds and pesticides controlling invertebrates. The monotypic habitat occurs in botanical and zoological contexts, and is a component of conservation biology. In restoration ecology of native plant communities or habitats, some invasive species create monotypic stands that replace and/or prevent other species, especially indigenous ones, from growing there. A dominant colonization can occur from retardant chemicals exuded, nutrient monopolization, or from lack of natural controls such as herbivores or climate, that keep them in balance with their native habitats. The yellow starthistle, *Centaurea solstitialis*, is a botanical monotypic-habitat example of this, currently dominating over 15,000,000 acres (61,000 km^2) in California alone. The non-native freshwater zebra mussel, *Dreissena polymorpha*, that colonizes areas of the Great Lakes and the Mississippi River watershed, is a zoological monotypic-habitat example; the predators that control it in its home-range in Russia are absent and it proliferates abundantly. Even though its name may seem to imply simplicity as compared with polytypic habitats, the monotypic habitat can be complex. Aquatic habitats, such as exotic *Hydrilla* beds, support a similarly rich fauna of macroinvertebrates to a more varied habitat, but the creatures present may differ between the two, affecting small fish and other animals higher up the food chainReferences

References

- Sutherland, William J.; Hill, David A. (1995). Managing Habitats for Conservation. Cambridge University Press. p. 6. ISBN 978-0-521-44776-8

- "A hydrothermal vent forms when seawater meets hot magma". Ocean facts. National Ocean Service. 11 January 2013. Retrieved 20 May 2016

- Abe, Y.; Bignell, David Edward; Higashi, T. (2014). Termites: Evolution, Sociality, Symbioses, Ecology. Springer. p. 437. ISBN 978-94-017-3223-9

- Richards, O.W. (1940). "The biology of the small white butterfly (Pieris rapae), with special reference to the factors controlling its abundance". Journal of Animal Ecology. 9 (2): 243–288. doi:10.2307/1459

- Hsing, Pen-Yuan (18 October 2010). "Gas-powered Circle of Life: Succession in a Deep-sea Ecosystem". Lophelia II 2010. NOAA. Retrieved 22 May 2016

- John G. Kelcey, John G. (2015). Vertebrates and Invertebrates of European Cities:Selected Non-Avian Fauna. Springer. p. 124. ISBN 978-1-4939-1698-6

Types of Habitats

There are various types of habitats that contain different species of flora and fauna. Habitats are unique and suited to the organisms that make them their home. Some types of habitats include marine habitats, deserts, etc. The topics discussed in the chapter are of great importance to broaden the existing knowledge on habitats.

Marine Habitats

The marine environment supplies many kinds of habitats that support marine life. Marine life depends in some way on the saltwater that is in the sea (the term *marine* comes from the Latin *mare*, meaning sea or ocean). A habitat is an ecological or environmental area inhabited by one or more living species.

Marine habitats can be divided into coastal and open ocean habitats. Coastal habitats are found in the area that extends from as far as the tide comes in on the shoreline out to the edge of the continental shelf. Most marine life is found in coastal habitats, even though the shelf area occupies only seven percent of the total ocean area. Open ocean habitats are found in the deep ocean beyond the edge of the continental shelf.

Alternatively, marine habitats can be divided into pelagic and demersal zones. Pelagic habitats are found near the surface or in the open water column, away from the bottom of the ocean. Demersal habitats are near or on the bottom of the ocean. An organism living in a pelagic habitat is said to be a pelagic organism, as in pelagic fish. Similarly, an organism living in a demersal habitat is said to be a demersal organism, as in demersal fish. Pelagic habitats are intrinsically shifting and ephemeral, depending on what ocean currents are doing.

Marine habitats can be modified by their inhabitants. Some marine organisms, like corals, kelp, mangroves and seagrasses, are ecosystem engineers which reshape the marine environment to the point where they create further habitat for other organisms.

Overview

In contrast to terrestrial habitats, marine habitats are shifting and ephemeral. Swimming organisms find areas by the edge of a continental shelf a good habitat, but only while upwellings bring nutrient rich water to the surface. Shellfish find habitat on sandy beaches, but storms, tides and currents mean their habitat continually reinvents itself.

The presence of seawater is common to all marine habitats. Beyond that many other things determine whether a marine area makes a good habitat and the type of habitat it makes. For example:

- temperature – is affected by geographical latitude, ocean currents, weather, the discharge of rivers, and by the presence of hydrothermal vents or cold seeps

- sunlight – photosynthetic processes depend on how deep and turbid the water is

- nutrients – are transported by ocean currents to different marine habitats from land run-off, or by upwellings from the deep sea, or they sink though the sea as marine snow

- salinity – varies, particularly in estuaries or near river deltas, or by hydrothermal vents

- dissolved gases – oxygen levels in particular, can be increased by wave actions and decreased during algal blooms

- acidity – this is partly to do with dissolved gases above, since the acidity of the ocean is largely controlled by how much carbon dioxide is in the water.

- turbulence – ocean waves, fast currents and the agitation of water affect the nature of habitats

- cover – the availability of cover such as the adjacency of the sea bottom, or the presence of floating objects

- the occupying organisms themselves – since organisms modify their habitats by the act of occupying them, and some, like corals, kelp, mangroves and seagrasses, create further habitats for other organisms.

Only 29 percent of the world surface is land. The rest is ocean, home to the marine habitats. The oceans are nearly four kilometres deep on average and are fringed with coastlines that run for nearly 380,000 kilometres.

Ocean	Area million km²	%	Volume million cu km	%	Mean depth km	Max depth km	Coastline km
Pacific Ocean	155.6	46.4	679.6	49.6	4.37	10.924	135,663
Atlantic Ocean	76.8	22.9	313.4	22.5	4.08	8.605	111,866
Indian Ocean	68.6	20.4	269.3	19.6	3.93	7.258	66,526
Southern Ocean	20.3	6.1	91.5	6.7	4.51	7.235	17,968
Arctic Ocean	14.1	4.2	17.0	1.2	1.21	4.665	45,389
Overall	335.3		1370.8		4.09	10.924	377,412

The ocean occupies 71 percent of the world surface, averaging nearly four kilometres in depth. There are five major oceans, of which the Pacific Ocean is nearly as large as the rest put together. Coastlines fringe the land for nearly 380,000 kilometres.

Land runoff, pouring into the sea, can contain nutrients

Marine habitats can be broadly divided into pelagic and demersal habitats. Pelagic habitats are the habitats of the open water column, away from the bottom of the ocean. Demersal habitats are the habitats that are near or on the bottom of the ocean. An organism living in a pelagic habitat is said to be a pelagic organism, as in pelagic fish. Similarly, an organism living in a demersal habitat is said to be a demersal organism, as in demersal fish. Pelagic habitats are intrinsically ephemeral, depending on what ocean currents are doing.

The land-based ecosystem depends on topsoil and fresh water, while the marine ecosystem depends on dissolved nutrients washed down from the land.

Ocean deoxygenation poses a threat to marine habitats, due to the growth of low oxygen zones.

Ocean Currents

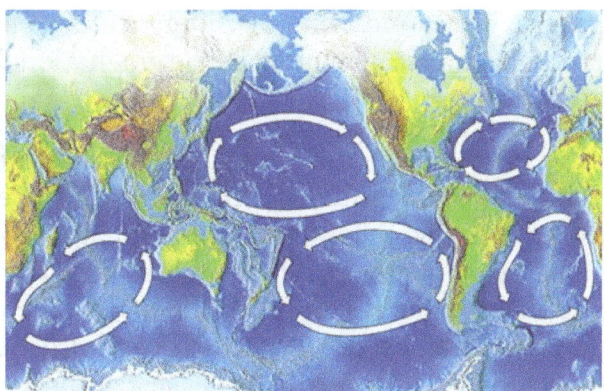

Ocean gyres rotate clockwise in the north and counterclockwise in the south.

In marine systems, ocean currents have a key role determining which areas are effective as habitats, since ocean currents transport the basic nutrients needed to support marine life. Plankton are the life forms that inhabit the ocean that are so small (less than 2 mm) that they cannot effectively propel themselves through the water, but must drift instead with the currents. If the current carries the right nutrients, and if it also flows at a suitably shallow depth where there is plenty of sunlight, then such a current itself can become a suitable habitat for photosynthesizing tiny algae called phytoplankton. These tiny plants are the primary producers in the ocean, at the start of the food chain. In turn, as the population of drifting phytoplankton grows, the water becomes a suitable habitat for zooplankton, which feed on the phytoplankton. While phytoplankton are tiny drifting plants, zooplankton are tiny drifting animals, such as the larvae of fish and marine invertebrates. If sufficient zooplankton establish themselves, the current becomes a candidate habitat for the forage fish that feed on them. And then if sufficient forage fish move to the area, it becomes a candidate habitat for larger predatory fish and other marine animals that feed on the forage fish. In this dynamic way, the current itself can, over time, become a moving habitat for multiple types of marine life.

This algae bloom occupies sunlit epipelagic waters off the southern coast of England.
The algae are maybe feeding on nutrients from land runoff or upwellings at the edge of the continental shelf.

Ocean currents can be generated by differences in the density of the water. How dense water is depends on how saline or warm it is. If water contains differences in salt content or temperature, then the different densities will initiate a current. Water that is saltier or cooler will be denser, and will sink in relation to the surrounding water. Conversely, warmer and less salty water will float to the surface. Atmospheric winds and pressure differences also produces surface currents, waves and seiches. Ocean currents are also generated by the gravitational pull of the sun and moon (tides), and seismic activity (tsunami).

The rotation of the Earth affects the direction ocean currents take, and explains which way the large circular ocean gyres rotate in the image above left. Suppose a current at the equator is heading north. The Earth rotates eastward, so the water possesses that rotational momentum. But the further the water moves north, the slower the earth moves eastward. If the current could get to the North Pole, the earth wouldn't be moving eastward at all. To conserve its rotational momentum, the further the current travels north the faster it must move eastward. So the effect is that the current curves to the right. This is the Coriolis effect. It is weakest at the equator and strongest at the poles. The effect is opposite south of the equator, where currents curve left.

Marine Topography

Marine (or seabed or ocean) topography refers to the shape the land has when it interfaces with the ocean. These shapes are obvious along coastlines, but they occur also in significant ways underwater. The effectiveness of marine habitats is partially defined by these shapes, including the way they interact with and shape ocean currents, and the way sunlight diminishes when these landforms occupy increasing depths. Tidal networks depend on the balance between sedimentary processes and hydrodynamics however, anthropogenic influences can impact the natural system more than any physical driver.

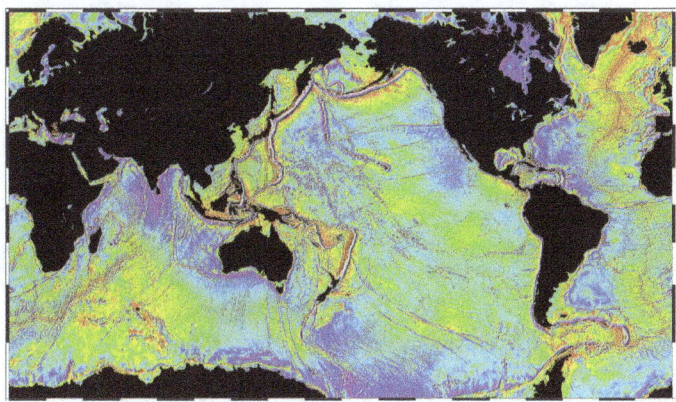

Map of underwater topography (1995 NOAA)

Marine topographies include coastal and oceanic landforms ranging from coastal estuaries and shorelines to continental shelves and coral reefs. Further out in the open ocean, they include underwater and deep sea features such as ocean rises and seamounts. The submerged surface has mountainous features, including a globe-spanning mid-ocean ridge system, as well as undersea volcanoes, oceanic trenches, submarine canyons, oceanic plateaus and abyssal plains.

The mass of the oceans is approximately 1.35×10^{18} metric tons, or about 1/4400 of the total mass of the Earth. The oceans cover an area of 3.618×10^{8} km² with a mean depth of 3,682 m, resulting in an estimated volume of 1.332×10^{9} km³.

Biomass

One measure of the relative importance of different marine habitats is the rate at which they produce biomass.

Producer	Biomass productivity (gC/m²/yr)	Total area (million km²)	Total production (billion tonnes C/yr)	Comment
swamps and marshes	2,500			Includes freshwater
coral reefs	2,000	0.28	0.56	
algal beds	2,000			
river estuaries	1,800			
open ocean	125	311	39	

Coastal

Marine coasts are dynamic environments which constantly change, like the ocean which partially shape them. The Earth's natural processes, including weather and sea level change, result in the erosion, accretion and resculpturing of coasts as well as the flooding and creation of continental shelves and drowned river valleys.

The main agents responsible for deposition and erosion along coastlines are waves, tides and currents. The formation of coasts also depends on the nature of the rocks they are made of – the harder the rocks the less likely they are to erode, so variations in rock hardness result in coastlines with different shapes.

Coastlines can be volatile habitats

Tides often determine the range over which sediment is deposited or eroded. Areas with high tidal ranges allow waves to reach farther up the shore, and areas with lower tidal ranges produce deposition at a smaller elevation interval. The tidal range is influenced by the size and shape of the coastline. Tides do not typically cause erosion by themselves; however, tidal bores can erode as the waves surge up river estuaries from the ocean.

Waves erode coastline as they break on shore releasing their energy; the larger the wave the more energy it releases and the more sediment it moves. Sediment deposited by waves comes from eroded cliff faces and is moved along the coastline by the waves. Sediment deposited by rivers is the dominant influence on the amount of sediment located on a coastline.

Shores that look permanent through the short perceptive of a human lifetime are in fact among the most temporary of all marine structures.

The sedimentologist Francis Shepard classified coasts as *primary* or *secondary*.

- Primary coasts are shaped by non-marine processes, by changes in the land form. If a coast is in much the same condition as it was when sea level was stabilised after the last ice age, it is called a primary coast. "Primary coasts are created by erosion (the wearing away of soil or rock), deposition (the buildup of sediment or sand) or tectonic activity (changes in the structure of the rock and soil because of earthquakes). Many of these coastlines were formed as the sea level rose during the last 18,000 years, submerging river and glacial valleys to form bays and fjords." An example of a primary coast is a river delta, which forms when a river deposits soil and other material as it enters the sea.

- Secondary coasts are produced by marine processes, such as the action of the sea or by creatures that live in it. Secondary coastlines include sea cliffs, barrier islands, mud flats, coral reefs, mangrove swamps and salt marshes.

Continental coastlines usually have a continental shelf, a shelf of relatively shallow water, less than 200 metres deep, which extends 68 km on average beyond the coast. Worldwide, continental shelves occupy a total area of about 24 million km² (9 million sq mi), 8% of the ocean's total area and nearly 5% of the world's total area. Since the continental shelf is usually less than 200 metres deep, it follows that coastal habitats are generally photic, situated in the sunlit epipelagic zone. This means the conditions for photosynthetic processes so important for primary production, are available to coastal marine habitats. Because land is nearby, there are large discharges of nutrient rich land runoff into coastal waters. Further, periodic upwellings from the deep ocean can provide cool and nutrient rich currents along the edge of the continental shelf.

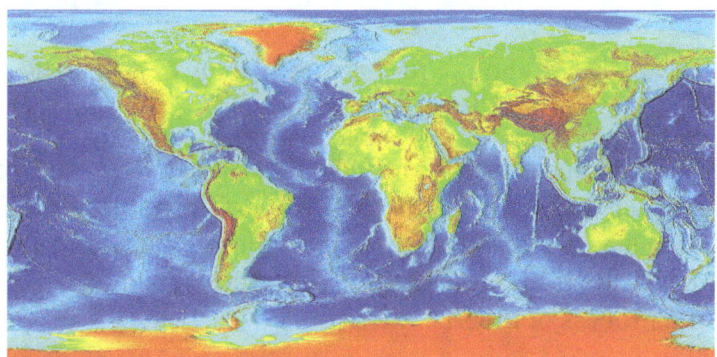

The global continental shelf, highlighted in cyan, defines the extent of coastal habitats, and occupies 5% of the total world area.

As a result, coastal marine life is the most abundant in the world. It is found in tidal pools, fjords and estuaries, near sandy shores and rocky coastlines, around coral reefs and on or above the continental shelf. Coastal fish include small forage fish as well as the larger predator fish that feed on them. Forage fish thrive in inshore waters where high productivity results from upwelling and shoreline run off of nutrients. Some are partial residents that spawn in streams, estuaries and bays, but most complete their life cycle in the zone. There can also be a mutualism between species that occupy adjacent marine habitats. For example, fringing reefs just below low tide level have a mutually beneficial relationship with mangrove forests at high tide level and sea grass meadows in between: the reefs protect the mangroves and seagrass from strong currents and waves that would damage them or erode the sediments in which they are rooted, while the mangroves and seagrass protect the coral from large influxes of silt, fresh water and pollutants. This additional level of variety in the environment is beneficial to many types of coral reef animals, which for example may feed in the sea grass and use the reefs for protection or breeding.

Waves and currents shape the intertidal shoreline, eroding the softer rocks and transporting and grading loose particles into shingles, sand or mud.

Coastal habitats are the most visible marine habitats, but they are not the only important marine habitats. Coastlines run for 380,000 kilometres, and the total volume of the ocean is 1,370 million cu km. This means that for each metre of coast, there is 3.6 cu km of ocean space available somewhere for marine habitats.

Intertidal

Intertidal zones, those areas close to shore, are constantly being exposed and covered by the ocean's tides. A huge array of life lives within this zone.

Shore habitats range from the upper intertidal zones to the area where land vegetation takes prominence. It can be underwater anywhere from daily to very infrequently. Many species here are scavengers, living off of sea life that is washed up on the shore. Many land animals also make much use of the shore and intertidal habitats. A subgroup of organisms in this habitat bores and grinds exposed rock through the process of bioerosion.

Sandy Shores

Sandy shores provide shifting homes to many species

Sandy shores, also called beaches, are coastal shorelines where sand accumulates. Waves and currents shift the sand, continually building and eroding the shoreline. Longshore currents flow parallel to the beaches, making waves break obliquely on the sand. These currents transport large amounts of sand along coasts, forming spits, barrier islands and tombolos. Longshore currents also commonly create offshore bars, which give beaches some stability by reducing erosion.

Sandy shores are full of life, The grains of sand host diatoms, bacteria and other microscopic creatures. Some fish and turtles return to certain beaches and spawn eggs in the sand. Birds habitat beaches, like gulls, loons, sandpipers, terns and pelicans. Aquatic mammals, such sea lions, recuperate on them. Clams, periwinkles, crabs, shrimp, starfish and sea urchins are found on most beaches.

Sand is a sediment made from small grains or particles with diameters between about 60 μm and 2 mm. Mud is a sediment made from particles finer than sand. This small particle size means that mud particles tend to stick together, whereas sand particles do not. Mud is not easily shifted by waves and currents, and when it dries out, cakes into a solid. By contrast, sand is easily shifted by waves and currents, and when sand dries out it can be blown in the wind, accumulating into shifting sand dunes. Beyond the high tide mark, if the beach is low-lying, the wind can form rolling hills of sand dunes. Small dunes shift and reshape under the influence of the wind while larger dunes stabilise the sand with vegetation.

Ocean processes grade loose sediments to particle sizes other than sand, such as gravel or cobbles. Waves breaking on a beach can leave a berm, which is a raised ridge of coarser pebbles or sand, at the high tide mark. Shingle beaches are made of particles larger than sand, such as cobbles, or small stones. These beaches make poor habitats. Little life survives because the stones are churned and pounded together by waves and currents.

Rocky Shores

Tidepools on rocky shores make turbulent habitats for many forms of marine life.

The relative solidity of rocky shores seems to give them a permanence compared to the shifting nature of sandy shores. This apparent stability is not real over even quite short geological time scales, but it is real enough over the short life of an organism. In contrast to sandy shores, plants and animals can anchor themselves to the rocks.

Competition can develop for the rocky spaces. For example, barnacles can compete successfully on open intertidal rock faces to the point where the rock surface is covered with them. Barnacles resist desiccation and grip well to exposed rock faces. However, in the crevices of the same rocks, the inhabitants are different. Here mussels can be the successful species, secured to the rock with their byssal threads.

Rocky and sandy coasts are vulnerable because humans find them attractive and want to live near them. An increasing proportion of the humans live by the coast, putting pressure on coastal habitats.

Mudflats

Mudflats are coastal wetlands that form when mud is deposited by tides or rivers. They are found in sheltered areas such as bays, bayous, lagoons, and estuaries. Mudflats may be viewed geologically as exposed layers of bay mud, resulting from deposition of estuarine silts, clays and marine animal detritus. Most of the sediment within a mudflat is within the intertidal zone, and thus the flat is submerged and exposed approximately twice daily.

Mudflats are typically important regions for wildlife, supporting a large population, although levels of biodiversity are not particularly high. They are of particular importance to migratory birds. In the United Kingdom mudflats have been classified as a Biodiversity Action Plan priority habitat.

Mudflats become temporary habitats for migrating birds

Mangrove and Salt Marshes

Mangroves provide nurseries for fish

Mangrove swamps and salt marshes form important coastal habitats in tropical and temperate areas respectively.

Mangroves are species of shrubs and medium size trees that grow in saline coastal sediment habitats in the tropics and subtropics – mainly between latitudes 25° N and 25° S. The saline conditions tolerated by various species range from brackish water, through pure seawater (30 to 40 ppt), to water concentrated by evaporation to over twice the salinity of ocean seawater (up to 90 ppt). There are many mangrove species, not all closely related. The term "mangrove" is used generally to cover all of these species, and it can be used narrowly to cover just mangrove trees of the genus *Rhizophora*.

Mangroves form a distinct characteristic saline woodland or shrubland habitat, called a *mangrove swamp* or *mangrove forest'*. Mangrove swamps are found in depositional coastal environments, where fine sediments (often with high organic content) collect in areas protected from high-energy wave action. Mangroves dominate three quarters of tropical coastlines.

Estuaries

Estuaries occur when rivers flow into a coastal bay or inlet. They are nutrient rich
and have a transition zone which moves from freshwater to saltwater.

An estuary is a partly enclosed coastal body of water with one or more rivers or streams flowing
into it, and with a free connection to the open sea. Estuaries form a transition zone between river
environments and ocean environments and are subject to both marine influences, such as tides,
waves, and the influx of saline water; and riverine influences, such as flows of fresh water and sediment. The inflow of both seawater and freshwater provide high levels of nutrients in both the water
column and sediment, making estuaries among the most productive natural habitats in the world.

Most estuaries were formed by the flooding of river-eroded or glacially scoured valleys when sea
level began to rise about 10,000-12,000 years ago. They are amongst the most heavily populated
areas throughout the world, with about 60% of the world's population living along estuaries and
the coast. As a result, estuaries are suffering degradation by many factors, including sedimentation
from soil erosion from deforestation; overgrazing and other poor farming practices; overfishing;
drainage and filling of wetlands; eutrophication due to excessive nutrients from sewage and animal wastes; pollutants including heavy metals, PCBs, radionuclides and hydrocarbons from sewage inputs; and diking or damming for flood control or water diversion.

Estuaries provide habitats for a large number of organisms and support very high productivity. Estuaries provide habitats for salmon and sea trout nurseries, as well as migratory bird
populations. Two of the main characteristics of estuarine life are the variability in salinity
and sedimentation. Many species of fish and invertebrates have various methods to control or
conform to the shifts in salt concentrations and are termed osmoconformers and osmoregula-
tors. Many animals also burrow to avoid predation and to live in the more stable sedimental
environment. However, large numbers of bacteria are found within the sediment which have
a very high oxygen demand. This reduces the levels of oxygen within the sediment often resulting in partially anoxic conditions, which can be further exacerbated by limited water flux.
Phytoplankton are key primary producers in estuaries. They move with the water bodies and
can be flushed in and out with the tides. Their productivity is largely dependent on the turbid-
ity of the water. The main phytoplankton present are diatoms and dinoflagellates which are
abundant in the sediment.

Kelp Forests

Kelp forests are underwater areas with a high density of kelp. They form some of the most productive and dynamic ecosystems on Earth. Smaller areas of anchored kelp are called *kelp beds*. Kelp forests occur worldwide throughout temperate and polar coastal oceans.

Kelp forests provide a unique three-dimensional habitat for marine organisms and are a source for understanding many ecological processes. Over the last century, they have been the focus of extensive research, particularly in trophic ecology, and continue to provoke important ideas that are relevant beyond this unique ecosystem. For example, kelp forests can influence coastal oceanographic patterns and provide many ecosystem services.

However, humans have contributed to kelp forest degradation. Of particular concern are the effects of overfishing nearshore ecosystems, which can release herbivores from their normal population regulation and result in the over-grazing of kelp and other algae. This can rapidly result in transitions to barren landscapes where relatively few species persist.

Frequently considered an ecosystem engineer, kelp provides a physical substrate and habitat for kelp forest communities. In algae (Kingdom: Protista), the body of an individual organism is known as a thallus rather than as a plant (Kingdom: Plantae). The morphological structure of a kelp thallus is defined by three basic structural units:

- The holdfast is a root-like mass that anchors the thallus to the sea floor, though unlike true roots it is not responsible for absorbing and delivering nutrients to the rest of the thallus;

- The stipe is analogous to a plant stalk, extending vertically from the holdfast and providing a support framework for other morphological features;

- The fronds are leaf- or blade-like attachments extending from the stipe, sometimes along its full length, and are the sites of nutrient uptake and photosynthetic activity.

Kelp forests provide habitat for many marine organisms

In addition, many kelp species have pneumatocysts, or gas-filled bladders, usually located at the base of fronds near the stipe. These structures provide the necessary buoyancy for kelp to maintain an upright position in the water column.

The environmental factors necessary for kelp to survive include hard substrate (usually rock), high nutrients (e.g., nitrogen, phosphorus), and light (minimum annual irradiance dose > 50 E m^{-2}). Especially productive kelp forests tend to be associated with areas of significant oceanographic upwelling, a process that delivers cool nutrient-rich water from depth to the ocean's mixed surface layer. Water flow and turbulence facilitate nutrient assimilation across kelp fronds throughout the water column. Water clarity affects the depth to which sufficient light can be transmitted. In ideal conditions, giant kelp (*Macrocystis spp.*) can grow as much as 30-60 centimetres vertically per day. Some species such as *Nereocystis* are annual while others like *Eisenia* are perennial, living for more than 20 years. In perennial kelp forests, maximum growth rates occur during upwelling months (typically spring and summer) and die-backs correspond to reduced nutrient availability, shorter photoperiods and increased storm frequency.

Seagrass Meadows

White-spotted puffers like living in seagrass areas

Seagrasses are flowering plants from one of four plant families which grow in marine environments. They are called *seagrasses* because the leaves are long and narrow and are very often green, and because the plants often grow in large meadows which look like grassland. Since seagrasses photosynthesize and are submerged, they must grow submerged in the photic zone, where there is enough sunlight. For this reason, most occur in shallow and sheltered coastal waters anchored in sand or mud bottoms.

Seagrasses form extensive beds or meadows, which can be either monospecific (made up of one species) or multispecific (where more than one species co-exist). Seagrass beds make highly diverse and productive ecosystems. They are home to phyla such as juvenile and adult fish, epiphytic and free-living macroalgae and microalgae, mollusks, bristle worms, and nematodes. Few species were originally considered to feed directly on seagrass leaves (partly because of their low nutritional content), but scientific reviews and improved working methods have shown that seagrass herbivory is a highly important link in the food chain, with hundreds of species feeding on seagrasses worldwide, including green turtles, dugongs, manatees, fish, geese, swans, sea urchins and crabs.

Seagrasses are ecosystem engineers in the sense that they partly create their own habitat. The leaves slow down water-currents increasing sedimentation, and the seagrass roots and rhizomes stabilize the seabed. Their importance to associated species is mainly due to provision of shelter (through their three-dimensional structure in the water column), and due to their extraordinarily high rate of primary production. As a result, seagrasses provide coastal zones with ecosystem services, such as fishing grounds, wave protection, oxygen production and protection against coastal erosion. Seagrass meadows account for 15% of the ocean's total carbon storage.

Coral Reefs

Reefs comprise some of the densest and most diverse habitats in the world. The best-known types of reefs are tropical coral reefs which exist in most tropical waters; however, reefs can also exist in cold water. Reefs are built up by corals and other calcium-depositing animals, usually on top of a rocky outcrop on the ocean floor. Reefs can also grow on other surfaces, which has made it possible to create artificial reefs. Coral reefs also support a huge community of life, including the corals themselves, their symbiotic zooxanthellae, tropical fish and many other organisms.

Much attention in marine biology is focused on coral reefs and the El Niño weather phenomenon. In 1998, coral reefs experienced the most severe mass bleaching events on record, when vast expanses of reefs across the world died because sea surface temperatures rose well above normal. Some reefs are recovering, but scientists say that between 50% and 70% of the world's coral reefs are now endangered and predict that global warming could exacerbate this trend.

Open Ocean

The open ocean is relatively unproductive because of a lack of nutrients, yet because it is so vast, it has more overall primary production than any other marine habitat. Only about 10 percent of marine species live in the open ocean. But among them are the largest and fastest of all marine animals, as well as the animals that dive the deepest and migrate the longest. In the depths lurk animal that, to our eyes, appear hugely alien.

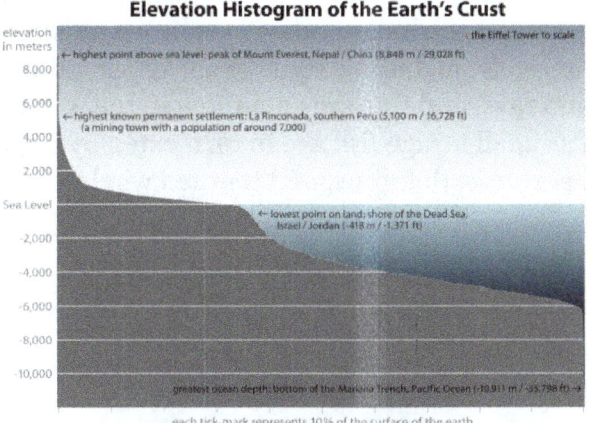

Elevation-area graph showing the proportion of land area at given heights and
the proportion of ocean area at given depths.

Surface Waters

In the open ocean, sunlit surface epipelagic waters get enough light for photosynthesis,
but there are often not enough nutrients. As a result, large areas contain little life apart from migrating animals.

The surface waters are sunlit. The waters down to about 200 metres are said to be in the epipelagic zone. Enough sunlight enters the epipelagic zone to allow photosynthesis by phytoplankton. The epipelagic zone is usually low in nutrients. This partially because the organic debris produced in the zone, such as excrement and dead animals, sink to the depths and are lost to the upper zone. Photosynthesis can happen only if both sunlight and nutrients are present.

In some places, like at the edge of continental shelves, nutrients can upwell from the ocean depth, or land runoff can be distributed by storms and ocean currents. In these areas, given that both sunlight and nutrients are now present, phytoplankton can rapidly establish itself, multiplying so fast that the water turns green from the chlorophyll, resulting in an algal bloom. These nutrient rich surface waters are among the most biologically productive in the world, supporting billions of tonnes of biomass.

"Phytoplankton are eaten by zooplankton - small animals which, like phytoplankton, drift in the ocean currents. The most abundant zooplankton species are copepods and krill: tiny crustaceans that are the most numerous animals on Earth. Other types of zooplankton include jelly fish and the larvae of fish, marine worms, starfish, and other marine organisms". In turn, the zooplankton are eaten by filter-feeding animals, including some seabirds, small forage fish like herrings and sardines, whale sharks, manta rays, and the largest animal in the world, the blue whale. Yet again, moving up the foodchain, the small forage fish are in turn eaten by larger predators, such as tuna, marlin, sharks, large squid, seabirds, dolphins, and toothed whales.

Deep Sea

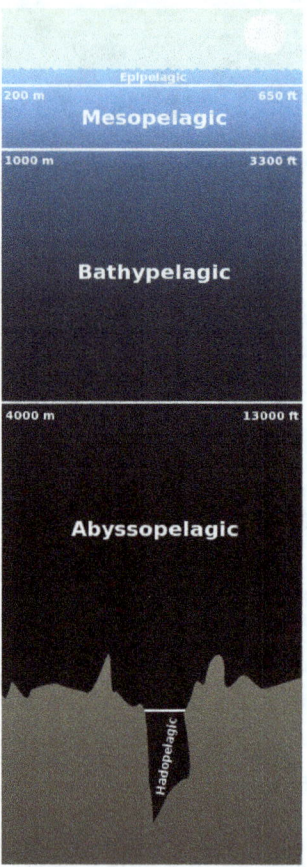

Scale diagram of the layers of the pelagic zone

The deep sea starts at the aphotic zone, the point where sunlight loses most of its energy in the water. Many life forms that live at these depths have the ability to create their own light a unique evolution known as bio-luminescence.

In the deep ocean, the waters extend far below the epipelagic zone, and support very different types of pelagic life forms adapted to living in these deeper zones.

Much of the aphotic zone's energy is supplied by the open ocean in the form of detritus. In deep water, marine snow is a continuous shower of mostly organic detritus falling from the upper layers of the water column. Its origin lies in activities within the productive photic zone. Marine snow includes dead or dying plankton, protists (diatoms), fecal matter, sand, soot and other inorganic dust. The "snowflakes" grow over time and may reach several centimetres in diameter, travelling for weeks before reaching the ocean floor. However, most organic components of marine snow are consumed by microbes, zooplankton and other filter-feeding animals within the first 1,000 metres of their journey, that is, within the epipelagic zone. In this way marine snow may be considered the foundation of deep-sea mesopelagic and benthic ecosystems: As sunlight cannot reach them, deep-sea organisms rely heavily on marine snow as an energy source.

Some deep-sea pelagic groups, such as the lanternfish, ridgehead, marine hatchetfish, and lightfish families are sometimes termed *pseudoceanic* because, rather than having an even distribution in open water, they occur in significantly higher abundances around structural oases, notably seamounts and over continental slopes. The phenomenon is explained by the likewise abundance of prey species which are also attracted to the structures.

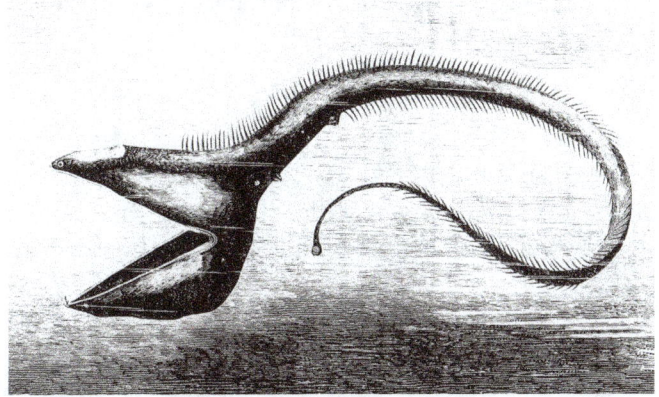

The umbrella mouth gulper eel can swallow a fish much larger than itself

The fish in the different pelagic and deep water benthic zones are physically structured, and behave in ways, that differ markedly from each other. Groups of coexisting species within each zone all seem to operate in similar ways, such as the small mesopelagic vertically migrating plankton-feeders, the bathypelagic anglerfishes, and the deep water benthic rattails. "

Ray finned species, with spiny fins, are rare among deep sea fishes, which suggests that deep sea fish are ancient and so well adapted to their environment that invasions by more modern fishes have been unsuccessful. The few ray fins that do exist are mainly in the Beryciformes and Lampriformes, which are also ancient forms. Most deep sea pelagic fishes belong to their own orders, suggesting a long evolution in deep sea environments. In contrast, deep water benthic species, are in orders that include many related shallow water fishes.

The umbrella mouth gulper is a deep sea eel with an enormous loosely hinged mouth. It can open its mouth wide enough to swallow a fish much larger than itself, and then expand its stomach to accommodate its catch.

Sea Floor

Vents and Seeps

Hydrothermal vents along the mid-ocean ridge spreading centers act as oases, as do their opposites, cold seeps. Such places support unique marine biomes and many new marine microorganisms and other lifeforms have been discovered at these locations.

Zooarium chimney provides a habitat for vent biota

Trenches

The deepest recorded oceanic trenches measure to date is the Mariana Trench, near the Philippines, in the Pacific Ocean at 10,924 m (35,838 ft). At such depths, water pressure is extreme and there is no sunlight, but some life still exists. A white flatfish, a shrimp and a jellyfish were seen by the American crew of the bathyscaphe *Trieste* when it dove to the bottom in 1960.

Seamounts

Marine life also flourishes around seamounts that rise from the depths, where fish and other sea life congregate to spawn and feed.

Coral Reef

Coral reefs are diverse underwater ecosystems held together by calcium carbonate structures secreted by corals. Coral reefs are built by colonies of tiny animals found in marine water that contain few nutrients. Most coral reefs are built from stony corals, which in turn consist of polyps that cluster in groups. The polyps belong to a group of animals known as Cnidaria, which also includes sea anemones and jellyfish. Unlike sea anemones, corals secrete hard carbonate exoskeletons which support and protect the coral polyps. Most reefs grow best in warm, shallow, clear, sunny and agitated water.

Often called "rainforests of the sea", shallow coral reefs form some of the most diverse ecosystems on Earth. They occupy less than 0.1% of the world's ocean surface, about half the area of France, yet they provide a home for at least 25% of all marine species, including fish, mollusks, worms, crustaceans, echinoderms, sponges, tunicates and other cnidarians. Paradoxically, coral reefs flourish even though they are surrounded by ocean waters that provide few nutrients. They are most commonly found at shallow depths in tropical waters, but deep water and cold water corals also exist on smaller scales in other areas.

Coral reefs deliver ecosystem services to tourism, fisheries and shoreline protection. The annual global economic value of coral reefs is estimated between US$29.8-375 billion. However, coral reefs are fragile ecosystems, partly because they are very sensitive to water temperature. They are under threat from climate change, oceanic acidification, blast fishing, cyanide fishing for aquarium fish, sunscreen use, overuse of reef resources, and harmful land-use practices, including urban and agricultural runoff and water pollution, which can harm reefs by encouraging excess algal growth.

Formation

Most of the coral reefs we can see today were formed after the last glacial period when melting ice caused the sea level to rise and flood the continental shelves. This means that most modern coral reefs are less than 10,000 years old. As communities established themselves on the shelves, the reefs grew upwards, pacing rising sea levels. Reefs that rose too slowly could become drowned reefs. They are covered by so much water that there was insufficient light. Coral reefs are found in the deep sea away from continental shelves, around oceanic islands and as atolls. The vast majority of these islands are volcanic in origin. The few exceptions have tectonic origins where plate movements have lifted the deep ocean floor on the surface.

In 1842 in his first monograph, *The Structure and Distribution of Coral Reefs*, Charles Darwin set out his theory of the formation of atoll reefs, an idea he conceived during the voyage of the *Beagle*. He theorized uplift and subsidence of the Earth's crust under the oceans formed the atolls. Darwin's theory sets out a sequence of three stages in atoll formation. It starts with a fringing reef forming around an extinct volcanic island as the island and ocean floor subsides. As the subsidence continues, the fringing reef becomes a barrier reef, and ultimately an atoll reef.

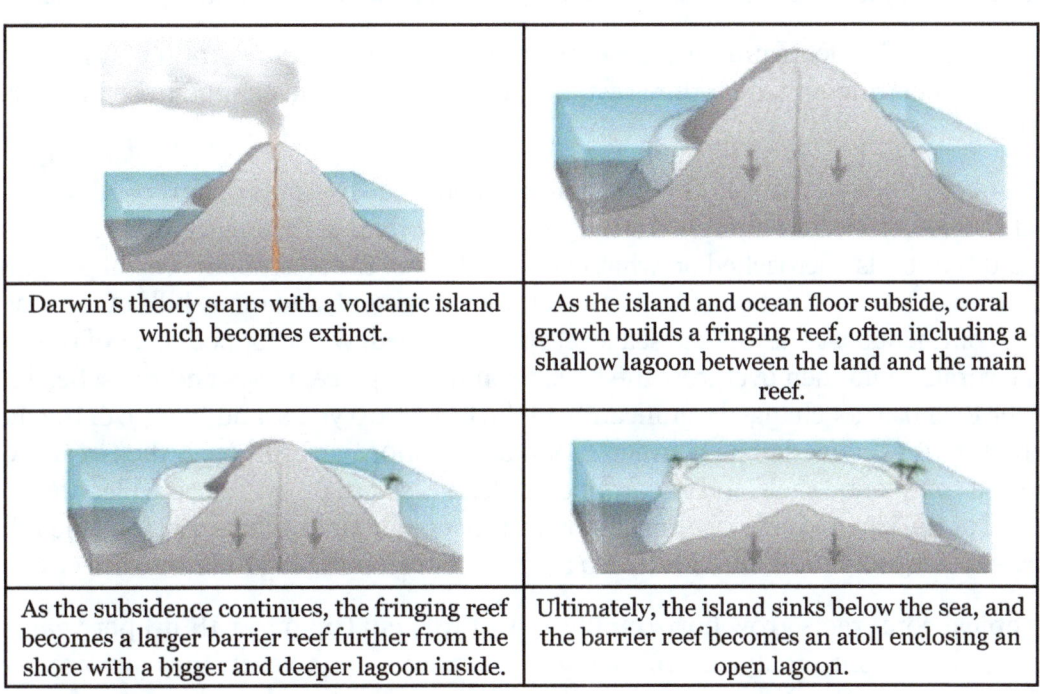

Darwin's theory starts with a volcanic island which becomes extinct.	As the island and ocean floor subside, coral growth builds a fringing reef, often including a shallow lagoon between the land and the main reef.
As the subsidence continues, the fringing reef becomes a larger barrier reef further from the shore with a bigger and deeper lagoon inside.	Ultimately, the island sinks below the sea, and the barrier reef becomes an atoll enclosing an open lagoon.

Darwin predicted that underneath each lagoon would be a bed rock base, the remains of the original volcano. Subsequent drilling proved this correct. Darwin's theory followed from his understanding that coral polyps thrive in the clean seas of the tropics where the water is agitated, but can only live within a limited depth range, starting just below low tide. Where the level of the underlying earth allows, the corals grow around the coast to form what he called fringing reefs, and can eventually grow out from the shore to become a barrier reef.

A fringing reef can take ten thousand years to form, and an atoll can take up to 30 million years.

Where the bottom is rising, fringing reefs can grow around the coast, but coral raised above sea level dies and becomes white limestone. If the land subsides slowly, the fringing reefs keep pace by growing upwards on a base of older, dead coral, forming a barrier reef enclosing a lagoon between the reef and the land. A barrier reef can encircle an island, and once the island sinks below sea level a roughly circular atoll of growing coral continues to keep up with the sea level, forming a central lagoon. Barrier reefs and atolls do not usually form complete circles, but are broken in places by storms. Like sea level rise, a rapidly subsiding bottom can overwhelm coral growth, killing the coral polyps and the reef, due to what is called *coral drowning*. Corals that rely on zooxanthellae can *drown* when the water becomes too deep for their symbionts to adequately photosynthesize, due to decreased light exposure.

The two main variables determining the geomorphology, or shape, of coral reefs are the nature of the underlying substrate on which they rest, and the history of the change in sea level relative to that substrate.

The approximately 20,000-year-old Great Barrier Reef offers an example of how coral reefs formed on continental shelves. Sea level was then 120 m (390 ft) lower than in the 21st century. As sea level rose, the water and the corals encroached on what had been hills of the Australian coastal plain. By 13,000 years ago, sea level had risen to 60 m (200 ft) lower than at present, and many hills of the coastal plains had become continental islands. As the sea level rise continued, water topped most of the continental islands. The corals could then overgrow the hills, forming the present cays and reefs. Sea level on the Great Barrier Reef has not changed significantly in the last 6,000 years, and the age of the modern living reef structure is estimated to be between 6,000 and 8,000 years. Although the Great Barrier Reef formed along a continental shelf, and not around a volcanic island, Darwin's principles apply. Development stopped at the barrier reef stage, since Australia is not about to submerge. It formed the world's largest barrier reef, 300–1,000 m (980–3,280 ft) from shore, stretching for 2,000 km (1,200 mi).

Healthy tropical coral reefs grow horizontally from 1 to 3 cm (0.39 to 1.18 in) per year, and grow vertically anywhere from 1 to 25 cm (0.39 to 9.84 in) per year; however, they grow only at depths shallower than 150 m (490 ft) because of their need for sunlight, and cannot grow above sea level.

Materials

As the name implies, the bulk of coral reefs is made up of coral skeletons from mostly intact coral colonies. As other chemical elements present in corals become incorporated into the calcium carbonate deposits, aragonite is formed. However, shell fragments and the remains of calcareous algae such as the green-segmented genus *Halimeda* can add to the reef's ability to withstand damage from storms and other threats. Such mixtures are visible in structures such as Eniwetok Atoll.

Types

The three principal reef types are:

- Fringing reef – directly attached to a shore, or borders it with an intervening shallow channel or lagoon

- Barrier reef – reef separated from a mainland or island shore by a deep channel or lagoon

- Atoll reef – more or less circular or continuous barrier reef extends all the way around a lagoon without a central island

A small atoll in the Maldives

Inhabited cay in the Maldives

Other reef types or variants are:

- Patch reef – common, isolated, comparatively small reef outcrop, usually within a lagoon or embayment, often circular and surrounded by sand or seagrass

- Apron reef – short reef resembling a fringing reef, but more sloped; extending out and downward from a point or peninsular shore

- Bank reef – linear or semicircular shaped-outline, larger than a patch reef

- Ribbon reef – long, narrow, possibly winding reef, usually associated with an atoll lagoon

- Table reef – isolated reef, approaching an atoll type, but without a lagoon

- Habili – reef specific to the Red Sea; does not reach the surface near enough to cause visible surf; may be a hazard to ships (from the Arabic for "unborn")

- Microatoll – community of species of corals; vertical growth limited by average tidal height; growth morphologies offer a low-resolution record of patterns of sea level change; fossilized remains can be dated using radioactive carbon dating and have been used to reconstruct Holocene sea levels

- Cays – small, low-elevation, sandy islands formed on the surface of coral reefs from eroded material that piles up, forming an area above sea level; can be stabilized by plants to become habitable; occur in tropical environments throughout the Pacific, Atlantic and Indian Oceans (including the Caribbean and on the Great Barrier Reef and Belize Barrier Reef), where they provide habitable and agricultural land

- Seamount or guyot – formed when a coral reef on a volcanic island subsides; tops of seamounts are rounded and guyots are flat; flat tops of guyots, or *tablemounts*, are due to erosion by waves, winds, and atmospheric processes

Zones

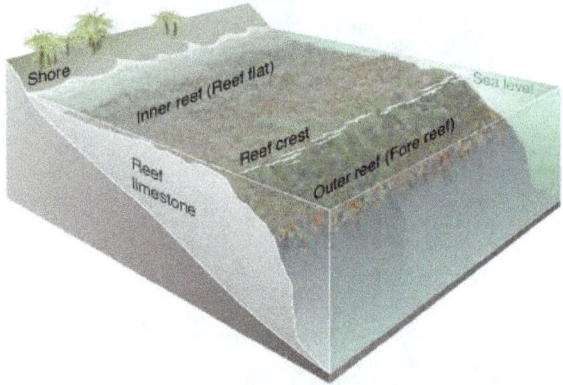

The three major zones of a coral reef: the fore reef, reef crest, and the back reef.

Coral reef ecosystems contain distinct zones that represent different kinds of habitats. Usually, three major zones are recognized: the fore reef, reef crest, and the back reef (frequently referred to as the reef lagoon).

All three zones are physically and ecologically interconnected. Reef life and oceanic processes create opportunities for exchange of seawater, sediments, nutrients, and marine life among one another.

Thus, they are integrated components of the coral reef ecosystem, each playing a role in the support of the reefs' abundant and diverse fish assemblages.

Most coral reefs exist in shallow waters less than 50 m deep. Some inhabit tropical continental shelves where cool, nutrient rich upwelling does not occur, such as Great Barrier Reef. Others are found in the deep ocean surrounding islands or as atolls, such as in the Maldives. The reefs surrounding islands form when islands subside into the ocean, and atolls form when an island subsides below the surface of the sea.

Alternatively, Moyle and Cech distinguish six zones, though most reefs possess only some of the zones.

Water in the reef surface zone is often agitated. This diagram represents a reef on a continental shelf. The water waves at the left travel over the *off-reef floor* until they encounter the *reef slope* or *fore reef*. Then the waves pass over the shallow *reef crest*. When a wave enters shallow water it shoals, that is, it slows down and the wave height increases.

The reef surface is the shallowest part of the reef. It is subject to the surge and the rise and fall of tides. When waves pass over shallow areas, they shoal, as shown in the diagram. This means the water is often agitated. These are the precise condition under which corals flourish. Shallowness means there is plenty of light for photosynthesis by the symbiotic zooxanthellae, and agitated water promotes the ability of coral to feed on plankton. However, other organisms must be able to withstand the robust conditions to flourish in this zone.

The off-reef floor is the shallow sea floor surrounding a reef. This zone occurs by reefs on continental shelves. Reefs around tropical islands and atolls drop abruptly to great depths, and do not have a floor. Usually sandy, the floor often supports seagrass meadows which are important foraging areas for reef fish.

The reef drop-off is, for its first 50 m, habitat for many reef fish who find shelter on the cliff face and plankton in the water nearby. The drop-off zone applies mainly to the reefs surrounding oceanic islands and atolls.

The reef face is the zone above the reef floor or the reef drop-off. This zone is often the most diverse area of the reef. Coral and calcareous algae growths provide complex habitats and areas which offer protection, such as cracks and crevices. Invertebrates and epiphytic algae provide much of the food for other organisms. A common feature on this forereef zone is spur and groove formations which serve to transport sediment downslope.

The reef flat is the sandy-bottomed flat, which can be behind the main reef, containing chunks of coral. This zone may border a lagoon and serve as a protective area, or it may lie between the reef and the shore, and in this case is a flat, rocky area. Fishes tend to prefer living in that flat, rocky area, compared to any other zone, when it is present.

The reef lagoon is an entirely enclosed region, which creates an area less affected by wave action that often contains small reef patches.

However, the "topography of coral reefs is constantly changing. Each reef is made up of irregular patches of algae, sessile invertebrates, and bare rock and sand. The size, shape and relative abundance of these patches changes from year to year in response to the various factors that favor one type of patch over another. Growing coral, for example, produces constant change in the fine structure of reefs. On a larger scale, tropical storms may knock out large sections of reef and cause boulders on sandy areas to move."

Locations

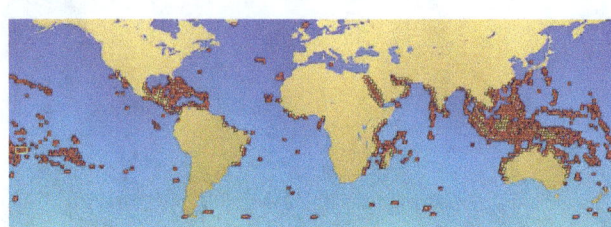

Locations of coral reefs

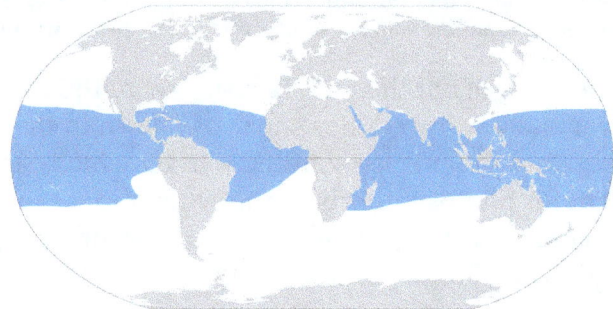

Boundary for 20 °C isotherms. Most corals live within this boundary.
Note the cooler waters caused by upwelling on the southwest coast of Africa and off the coast of Peru.

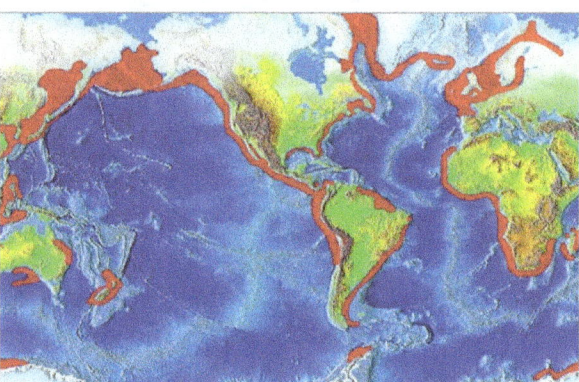

This map shows areas of upwelling in red. Coral reefs are not found in
coastal areas where colder and nutrient-rich upwellings occur.

Coral reefs are estimated to cover 284,300 km² (109,800 sq mi), just under 0.1% of the oceans' surface area. The Indo-Pacific region (including the Red Sea, Indian Ocean, Southeast Asia and the Pacific) account for 91.9% of this total. Southeast Asia accounts for 32.3% of that figure, while the Pacific including Australia accounts for 40.8%. Atlantic and Caribbean coral reefs account for 7.6%.

Although corals exist both in temperate and tropical waters, shallow-water reefs form only in a zone extending from approximately 30° N to 30° S of the equator. Tropical corals do not grow at depths of over 50 meters (160 ft). The optimum temperature for most coral reefs is 26–27 °C (79–81 °F), and few reefs exist in waters below 18 °C (64 °F). However, reefs in the Persian Gulf have adapted to temperatures of 13 °C (55 °F) in winter and 38 °C (100 °F) in summer. There are 37 species of scleractinian corals identified in such harsh environment around Larak Island.

Marine biodiversity of Raja Ampat Islands, Indonesia. It is estimated that about 75% of world coral reef population are in the Raja Ampat Islands.

Deep-water coral can exist at greater depths and colder temperatures at much higher latitudes, as far north as Norway. Although deep water corals can form reefs, very little is known about them.

Coral reefs are rare along the west coasts of the Americas and Africa, due primarily to upwelling and strong cold coastal currents that reduce water temperatures in these areas (respectively the Peru, Benguela and Canary streams). Corals are seldom found along the coastline of South Asia—from the eastern tip of India (Chennai) to the Bangladesh and Myanmar borders—as well as along the coasts of northeastern South America and Bangladesh, due to the freshwater release from the Amazon and Ganges Rivers respectively.

- The Great Barrier Reef—largest, comprising over 2,900 individual reefs and 900 islands stretching for over 2,600 kilometers (1,600 mi) off Queensland, Australia.

- The Mesoamerican Barrier Reef System—second largest, stretching 1,000 kilometers (620 mi) from Isla Contoy at the tip of the Yucatán Peninsula down to the Bay Islands of Honduras.

- The New Caledonia Barrier Reef—second longest double barrier reef, covering 1,500 kilometers (930 mi).

- The Andros, Bahamas Barrier Reef—third largest, following the east coast of Andros Island, Bahamas, between Andros and Nassau.

- The Red Sea—includes 6000-year-old fringing reefs located around a 2,000 km (1,240 mi) coastline.

- The Florida Reef Tract—largest continental US reef, extends from Soldier Key, located in Biscayne Bay, to the Dry Tortugas in the Gulf of Mexico.

- Pulley Ridge—deepest photosynthetic coral reef, Florida.

- Numerous reefs scattered over the Maldives.

- The Philippines coral reef area, the second largest in Southeast Asia, is estimated at 26,000 square kilometers and holds an extraordinary diversity of species. Scientists have identified 915 reef fish species and more than 400 scleractinian coral species, 12 of which are endemic.

- The Raja Ampat Islands in Indonesia's West Papua province offer the highest known marine diversity.

- Bermuda is known for its northernmost coral reef system, located at 32.4° N and 64.8° W. The presence of coral reefs at this high latitude is due to the proximity of the Gulf Stream. Bermuda has a fairly consistent diversity of coral species, representing a subset of those found in the greater Caribbean.

- The world's northernmost individual coral reef so far discovered is located within a bay of Japan's Tsushima Island in the Korea Strait.

- The world's southernmost coral reef is at Lord Howe Island, in the Pacific Ocean off the east coast of Australia.

Biology

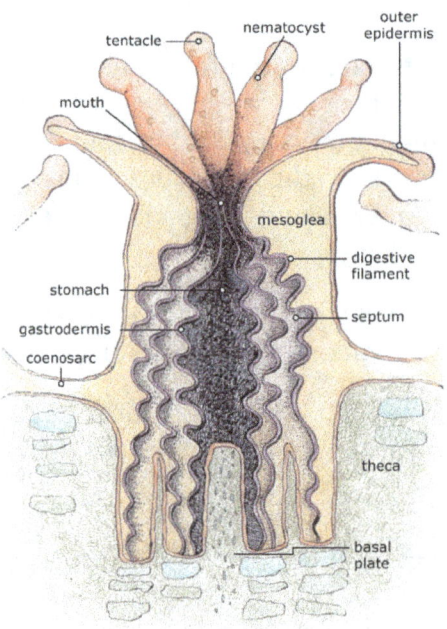

Anatomy of a coral polyp

Alive corals are colonies of small animals embedded in calcium carbonate shells. It is a mistake to think of coral as plants or rocks. Coral heads consist of accumulations of individual animals called polyps, arranged in diverse shapes. Polyps are usually tiny, but they can range in size from a pinhead to 12 inches (30 cm) across.

Reef-building or hermatypic corals live only in the photic zone (above 50 m), the depth to which sufficient sunlight penetrates the water, allowing photosynthesis to occur. Coral polyps do not photosynthesize, but have a symbiotic relationship with microscopic algae of the genus *Symbiodinium*, commonly referred to as zooxanthellae. These organisms live within the tissues of polyps and provide organic nutrients that nourish the polyp. Because of this relationship, coral reefs grow much faster in clear water, which admits more sunlight. Without their symbionts, coral growth would be too slow to form significant reef structures. Corals get up to 90% of their nutrients from their symbionts.

Reefs grow as polyps and other organisms deposit calcium carbonate, the basis of coral, as a skeletal structure beneath and around themselves, pushing the coral head's top upwards and outwards. Waves, grazing fish (such as parrotfish), sea urchins, sponges, and other forces and organisms act as bioeroders, breaking down coral skeletons into fragments that settle into spaces in the reef structure or form sandy bottoms in associated reef lagoons. Many other organisms living in the reef community contribute skeletal calcium carbonate in the same manner. Coralline algae are important contributors to reef structure in those parts of the reef subjected to the greatest forces by waves (such as the reef front facing the open ocean). These algae strengthen the reef structure by depositing limestone in sheets over the reef surface.

Typical shapes for coral species are wrinkled brains, cabbages, table tops, antlers, wire strands and pillars. These shapes can depend on the life history of the coral, like light exposure and wave action, and events such as breakages.

Table coral

Close up of polyps are arrayed on a coral, waving their tentacles.
There can be thousands of polyps on a single coral branch.

Corals reproduce both sexually and asexually. An individual polyp uses both reproductive modes within its lifetime. Corals reproduce sexually by either internal or external fertilization. The reproductive cells are found on the mesenteries, membranes that radiate inward from the layer of tissue that lines the stomach cavity. Some mature adult corals are hermaphroditic; others are exclusively male or female. A few species change sex as they grow.

Internally fertilized eggs develop in the polyp for a period ranging from days to weeks. Subsequent development produces a tiny larva, known as a planula. Externally fertilized eggs develop during synchronized spawning. Polyps release eggs and sperm into the water en masse, simultaneously. Eggs disperse over a large area. The timing of spawning depends on time of year, water temperature, and tidal and lunar cycles. Spawning is most successful when there is little variation between high and low tide. The less water movement, the better the chance for fertilization. Ideal timing occurs in the spring. Release of eggs or planula usually occurs at night, and is sometimes in phase with the lunar cycle (three to six days after a full moon). The period from release to settlement lasts only a few days, but some planulae can survive afloat for several weeks. They are vulnerable to predation and environmental conditions. The lucky few planulae which successfully attach to substrate next confront competition for food and space.

There are eight clades of Symbiodinium phylotypes. Most research has been completed on the Symbiodinium clades A–D. Each one of the eight contributes their own benefits as well as less compatible attributes to the survival of their coral hosts. Each photosynthetic organism has a specific level of sensitivity to photodamage of compounds needed for survival, such as proteins. Rates of regeneration and replication determine the organism's ability to survive. Phylotype A is found more in the shallow regions of marine waters. It is able to produce mycosporine-like amino acids that are UV resistant, using a derivative of glycerin to absorb the UV radiation and allowing them to become more receptive to warmer water temperatures. In the event of UV or thermal damage, if and when repair occurs, it will increase the likelihood of survival of the host and symbiont. This leads to the idea that, evolutionarily, clade A is more UV resistant and thermally resistant than the other clades.

Clades B and C are found more frequently in the deeper water regions, which may explain the higher susceptibility to increased temperatures. Terrestrial plants that receive less sunlight because they are found in the undergrowth can be analogized to clades B, C, and D. Since clades B through D are found at deeper depths, they require an elevated light absorption rate to be able to synthesize as much energy. With elevated absorption rates at UV wavelengths, the deeper occurring phylotypes are more prone to coral bleaching versus the more shallow clades. Clade D has been observed to be high temperature-tolerant, and as a result it has a higher rate of survival than clades B and C.

| Staghorn coral | Maze coral | Black coral | Fluorescent coral |

Darwin's Paradox

Darwin's paradox

"Coral... seems to proliferate when ocean waters are warm, poor, clear and agitated, a fact which

Darwin had already noted when he passed through Tahiti in 1842. This constitutes a fundamental paradox, shown quantitatively by the apparent impossibility of balancing input and output of the nutritive elements which control the coral polyp metabolism.

Recent oceanographic research has brought to light the reality of this paradox by confirming that the oligotrophy of the ocean euphotic zone persists right up to the swell-battered reef crest. When you approach the reef edges and atolls from the quasidesert of the open sea, the near absence of living matter suddenly becomes a plethora of life, without transition. So why is there something rather than nothing, and more precisely, where do the necessary nutrients for the functioning of this extraordinary coral reef machine come from?"

— Francis Rougerie

In *The Structure and Distribution of Coral Reefs*, published in 1842, Darwin described how coral reefs were found in some areas of the tropical seas but not others, with no obvious cause. The largest and strongest corals grew in parts of the reef exposed to the most violent surf and corals were weakened or absent where loose sediment accumulated.

Tropical waters contain few nutrients yet a coral reef can flourish like an "oasis in the desert". This has given rise to the ecosystem conundrum, sometimes called "Darwin's paradox": "How can such high production flourish in such nutrient poor conditions?"

Coral reefs cover less than 0.1% of the surface of the world's ocean, about half the land area of France, yet they support over one-quarter of all marine species. This diversity results in complex food webs, with large predator fish eating smaller forage fish that eat yet smaller zooplankton and so on. However, all food webs eventually depend on plants, which are the primary producers. Coral reefs' primary productivity is very high, typically producing 5–10 grams of carbon per square meter per day ($gC \cdot m^{-2} \cdot day^{-1}$) biomass.

Coral polyps

One reason for the unusual clarity of tropical waters is they are deficient in nutrients and drifting plankton. Further, the sun shines year-round in the tropics, warming the surface layer, making it less dense than subsurface layers. The warmer water is separated from deeper, cooler water by a stable thermocline, where the temperature makes a rapid change. This keeps the warm surface waters floating above the cooler deeper waters. In most parts of the ocean, there is little exchange between these layers. Organisms that die in aquatic environments generally sink to the bottom, where they decompose, which releases nutrients in the form of nitrogen (N), phosphorus (P) and potassium (K). These nutrients are necessary for plant growth, but in the tropics, they do not directly return to the surface.

Plants form the base of the food chain, and need sunlight and nutrients to grow. In the ocean, these plants are mainly microscopic phytoplankton which drift in the water column. They need sunlight for photosynthesis, which powers carbon fixation, so they are found only relatively near the surface. But they also need nutrients. Phytoplankton rapidly use nutrients in the surface waters, and in the tropics, these nutrients are not usually replaced because of the thermocline.

Explanations

Around coral reefs, lagoons fill in with material eroded from the reef and the island. They become havens for marine life, providing protection from waves and storms.

Most importantly, reefs recycle nutrients, which happens much less in the open ocean. In coral reefs and lagoons, producers include phytoplankton, as well as seaweed and coralline algae, especially small types called turf algae, which pass nutrients to corals. The phytoplankton are eaten by fish and crustaceans, who also pass nutrients along the food web. Recycling ensures fewer nutrients are needed overall to support the community.

Coral reefs support many symbiotic relationships. In particular, zooxanthellae provide energy to coral in the form of glucose, glycerol, and amino acids. Zooxanthellae can provide up to 90% of a coral's energy requirements. In return, as an example of mutualism, the corals shelter the zooxanthellae, averaging one million for every cubic centimeter of coral, and provide a constant supply of the carbon dioxide they need for photosynthesis.

The color of corals depends on the combination of brown shades provided by their
zooxanthellae and pigmented proteins (reds, blues, greens, etc.) produced by the corals themselves.

Corals also absorb nutrients, including inorganic nitrogen and phosphorus, directly from water. Many corals extend their tentacles at night to catch zooplankton that brush them when the water is agitated. Zooplankton provide the polyp with nitrogen, and the polyp shares some of the nitrogen with the zooxanthellae, which also require this element. The varying pigments in different species of zooxanthellae give them an overall brown or golden-brown appearance, and give brown corals their colors. Other pigments such as reds, blues, greens, etc. come from colored proteins made by the coral animals. Coral which loses a large fraction of its zooxanthellae becomes white (or sometimes pastel shades in corals that are richly pigmented with their own colorful proteins) and is said to be bleached, a condition which, unless corrected, can kill the coral.

Sponges are another key: they live in crevices in the coral reefs. They are efficient filter feeders, and in the Red Sea they consume about 60% of the phytoplankton that drifts by. The sponges eventually excrete nutrients in a form the corals can use.

Most coral polyps are nocturnal feeders. Here, in the dark, polyps
have extended their tentacles to feed on zooplankton.

The roughness of coral surfaces is the key to coral survival in agitated waters. Normally, a boundary layer of still water surrounds a submerged object, which acts as a barrier. Waves breaking on the extremely rough edges of corals disrupt the boundary layer, allowing the corals access to passing nutrients. Turbulent water thereby promotes reef growth and branching. Without the nutritional gains brought by rough coral surfaces, even the most effective recycling would leave corals wanting in nutrients.

Studies have shown that deep nutrient-rich water entering coral reefs through isolated events may have significant effects on temperature and nutrient systems. This water movement disrupts the relatively stable thermocline that usually exists between warm shallow water to deeper colder water. Leichter et al. (2006) found that temperature regimes on coral reefs in the Bahamas and Florida were highly variable with temporal scales of minutes to seasons and spatial scales across depths.

Water can be moved through coral reefs in various ways, including current rings, surface waves, internal waves and tidal changes. Movement is generally created by tides and wind. As tides interact with varying bathymetry and wind mixes with surface water, internal waves are created. An internal wave is a gravity wave that moves along density stratification within the ocean. When a water parcel encounters a different density it will oscillate and create internal waves. While internal waves generally have a lower frequency than surface waves, they often form as a single wave that breaks into multiple waves as it hits a slope and moves upward. This vertical break up of internal waves causes significant diapycnal mixing and turbulence. Internal waves can act as nutrient pumps, bringing plankton and cool nutrient-rich water up to the surface.

The irregular structure characteristic of coral reef bathymetry may enhance mixing and produce pockets of cooler water and variable nutrient content. Arrival of cool, nutrient-rich water from depths due to internal waves and tidal bores has been linked to growth rates of suspension feeders and benthic algae as well as plankton and larval organisms. Leichter et al. proposed that Codium isthmocladum react to deep water nutrient sources due to their tissues having different concentrations of nutrients dependent upon depth. Wolanski and Hamner noted aggregations of eggs, larval organisms and plankton on reefs in response to deep water intrusions. Similarly, as internal waves and bores move vertically, surface-dwelling larval organisms are carried toward the shore. This has significant biological importance to cascading effects of food chains in coral reef ecosystems and may provide yet another key to unlocking "Darwin's Paradox".

Cyanobacteria provide soluble nitrates for the reef via nitrogen fixation.

Coral reefs also often depend on surrounding habitats, such as seagrass meadows and mangrove forests, for nutrients. Seagrass and mangroves supply dead plants and animals which are rich in nitrogen and also serve to feed fish and animals from the reef by supplying wood and vegetation. Reefs, in turn, protect mangroves and seagrass from waves and produce sediment in which the mangroves and seagrass can root.

Biodiversity

Tube sponges attracting cardinal fishes, glassfishes and wrasses

Over 4,000 species of fish inhabit coral reefs

Coral reefs form some of the world's most productive ecosystems, providing complex and varied marine habitats that support a wide range of other organisms.Fringing reefs just below low tide level have a mutually beneficial relationship with mangrove forests at high tide level and sea grass meadows in between: the reefs protect the mangroves and seagrass from strong currents and waves that would damage them or erode the sediments in which they are rooted, while the mangroves and sea grass protect the coral from large influxes of silt, fresh water and pollutants. This level of variety in the environment benefits many coral reef animals, which, for example, may feed in the sea grass and use the reefs for protection or breeding.

Reefs are home to a large variety of animals, including fish, seabirds, sponges, cnidarians (which includes some types of corals and jellyfish), worms, crustaceans (including shrimp, cleaner shrimp, spiny lobsters and crabs), mollusks (including cephalopods), echinoderms (including starfish, sea urchins and sea cucumbers), sea squirts, sea turtles and sea snakes. Aside from humans, mammals are rare on coral reefs, with visiting cetaceans such as dolphins being the main exception. A few of these varied species feed directly on corals, while others graze on algae on the reef. Reef biomass is positively related to species diversity.

Organisms can cover every square inch of a coral reef

The same hideouts in a reef may be regularly inhabited by different species at different times of day. Nighttime predators such as cardinalfish and squirrelfish hide during the day, while damselfish, surgeonfish, triggerfish, wrasses and parrotfish hide from eels and sharks.

Algae

Reefs are chronically at risk of algal encroachment. Overfishing and excess nutrient supply from onshore can enable algae to outcompete and kill the coral. Increased nutrient levels can be a result of sewage or chemical fertilizer runoff from nearby coastal developments. Runoff can carry nitrogen and phosphorus which promote excess algae growth. Algae can sometimes out-compete the coral for space. The algae can then smother the coral by decreasing the oxygen supply available to the reef. Decreased oxygen levels can slow down coral's calcification rates weakening the coral and leaving it more susceptible to disease and degradation. In surveys done around largely uninhabited US Pacific islands, algae inhabit a large percentage of surveyed coral locations. The algal population consists of turf algae, coralline algae, and macro algae.

Sponges

Sponges are essential for the functioning of the coral reef's ecosystem. Algae and corals in coral reefs produce organic material. This is filtered through sponges which convert this organic material into small particles which in turn are absorbed by algae and corals.

Fish

Over 4,000 species of fish inhabit coral reefs. The reasons for this diversity remain unclear. Hypotheses include the "lottery", in which the first (lucky winner) recruit to a territory is typically able to defend it against latecomers, "competition", in which adults compete for territory, and less-competitive species must be able to survive in poorer habitat, and "predation", in which population size is a function of postsettlement piscivore mortality. Healthy reefs can produce up to 35 tons of fish per square kilometer each year, but damaged reefs produce much less.

Invertebrates

Sea urchins, Dotidae and sea slugs eat seaweed. Some species of sea urchins, such as *Diadema antillarum*, can play a pivotal part in preventing algae from overrunning reefs. Nudibranchia and sea anemones eat sponges.

A number of invertebrates, collectively called "cryptofauna," inhabit the coral skeletal substrate itself, either boring into the skeletons (through the process of bioerosion) or living in pre-existing voids and crevices. Those animals boring into the rock include sponges, bivalve mollusks, and sipunculans. Those settling on the reef include many other species, particularly crustaceans and polychaete worms.

Seabirds

Coral reef systems provide important habitats for seabird species, some endangered. For example, Midway Atoll in Hawaii supports nearly three million seabirds, including two-thirds (1.5 million) of the global population of Laysan albatross, and one-third of the global population of black-footed albatross. Each seabird species has specific sites on the atoll where they nest. Altogether, 17 species of seabirds live on Midway. The short-tailed albatross is the rarest, with fewer than 2,200 surviving after excessive feather hunting in the late 19th century.

Other

Sea snakes feed exclusively on fish and their eggs. Marine birds, such as herons, gannets, pelicans and boobies, feed on reef fish. Some land-based reptiles intermittently associate with reefs, such as monitor lizards, the marine crocodile and semiaquatic snakes, such as *Laticauda colubrina*. Sea turtles, particularly hawksbill sea turtles, feed on sponges.

Schooling reef fish	Caribbean reef squid
Whitetip reef shark	Green turtle
Giant clam	Soft coral, cup coral, sponges and ascidians

Importance

Coral reefs deliver ecosystem services to tourism, fisheries and coastline protection. The global economic value of coral reefs has been estimated to be between US $29.8 billion and $375 billion per year. Coral reefs protect shorelines by absorbing wave energy, and many small islands would not exist without their reefs to protect them. According to the environmental group World Wide Fund for Nature, the economic cost over a 25-year period of destroying one kilometer of coral reef is somewhere between $137,000 and $1,200,000. About six million tons of fish are taken each year from coral reefs. Well-managed coral reefs have an annual yield of 15 tons of seafood on average per square kilometer. Southeast Asia's coral reef fisheries alone yield about $2.4 billion annually from seafood.

To improve the management of coastal coral reefs, another environmental group, the World Resources Institute (WRI) developed and published tools for calculating the value of coral reef-related tourism, shoreline protection and fisheries, partnering with five Caribbean countries. As of April 2011, published working papers covered St. Lucia, Tobago, Belize, and the Dominican Republic, with a paper for Jamaica in preparation. The WRI was also "making sure that the study results support improved coastal policies and management planning". The Belize study estimated the value of reef and mangrove services at $395–559 million annually.

Bermuda's coral reefs provide economic benefits to the Island worth on average $722 million per year, based on six key ecosystem services, according to Sarkis *et al* (2010).

Threats

Island with fringing reef off Yap, Micronesia

Coral reefs are dying around the world. In particular, coral mining, agricultural and urban run-off, pollution (organic and inorganic), overfishing, blast fishing, disease, and the digging of canals and access into islands and bays are localized threats to coral ecosystems. Broader threats are sea temperature rise, sea level rise and pH changes from ocean acidification, all associated with green-house gas emissions. A 2014 study lists factors such as population explosion along the coast lines, overfishing, the pollution of coastal areas, global warming and invasive species among the main reasons that have put reefs in danger of extinction.

A study released in April 2013 has shown that air pollution can also stunt the growth of coral reefs; researchers from Australia, Panama and the UK used coral records (between 1880 and 2000) from the western Caribbean to show the threat of factors such as coal-burning coal and volcanic eruptions. Pollutants, such as Tributyltin, a biocide released into water from in anti-fouling paint can be toxic to corals.

In 2011, researchers suggested that "extant marine invertebrates face the same synergistic effects of multiple stressors" that occurred during the end-Permian extinction, and that genera "with poorly buffered respiratory physiology and calcareous shells", such as corals, were particularly vulnerable.

Rock coral on seamounts across the ocean are under fire from bottom trawling. Reportedly up to 50% of the catch is rock coral, and the practice transforms coral structures to rubble. With it taking years to regrow, these coral communities are disappearing faster than they can sustain themselves.

Another cause for the death of coral reefs is bioerosion. Various fishes graze corals, dead or alive and change the morphology of coral reefs making them more susceptible to other physical and chemical threats. It has been generally observed that only the algae growing on dead corals is eaten and the live ones are not. However, this act still destroys the top layer of coral substrate and makes it harder for the reefs to sustain.

In El Niño-year 2010, preliminary reports show global coral bleaching reached its worst level since another El Niño year, 1998, when 16% of the world's reefs died as a result of increased water temperature. In Indonesia's Aceh province, surveys showed some 80% of bleached corals died. Scientists do not yet understand the long-term impacts of coral bleaching, but they do know that bleaching leaves corals vulnerable to disease, stunts their growth, and affects their reproduction, while severe bleaching kills them. In July, Malaysia closed several dive sites where virtually all the corals were damaged by bleaching.

To find answers for these problems, researchers study the various factors that impact reefs. The list includes the ocean's role as a carbon dioxide sink, atmospheric changes, ultraviolet light, ocean acidification, viruses, impacts of dust storms carrying agents to far-flung reefs, pollutants, algal blooms and others. Reefs are threatened well beyond coastal areas. Coral reefs with one type of zooxanthellae are more prone to bleaching than are reefs with another, more hardy, species.

General estimates show approximately 10% of the world's coral reefs are dead. About 60% of the world's reefs are at risk due to destructive, human-related activities. The threat to the health of reefs is particularly high in Southeast Asia, where 95% of reefs are at risk from local threats. By the 2030s, 90% of reefs are expected to be at risk from both human activities and climate change; by 2050, all coral reefs will be in danger.

Current research is showing that ecotourism in the Great Barrier Reef is contributing to coral disease, and that chemicals in sunscreens may contribute to the impact of viruses on zooxanthellae.

Some scientists, including those associated with the National Oceanic and Atmospheric Administration, posit that US coral reefs are likely to disappear within a few decades as a result of global warming.

Protection

A diversity of corals

Marine protected areas (MPAs) have become increasingly prominent for reef management. MPAs promote responsible fishery management and habitat protection. Much like national parks and wildlife refuges, and to varying degrees, MPAs restrict potentially damaging activities. MPAs encompass both social and biological objectives, including reef restoration, aesthetics, biodiversity, and economic benefits. However, there are very few MPAs that have actually made a substantial difference. Research in Indonesia, Philippines and Papua New Guinea shows that there is no significant difference between an MPA site and an unprotected site. Conflicts surrounding MPAs involve lack of participation, clashing views of the government and fisheries, effectiveness of the area, and funding. In some situations, as in the Phoenix Islands Protected Area, MPAs can also provide revenue, potentially equal to the income they would have generated without controls, as Kiribati did for its Phoenix Islands.

According to the *Caribbean Coral Reefs - Status Report 1970-2012* made by the IUCN. States that; stopping overfishing especially key fishes to coral reef like parrotfish, coastal zone management which reduce human pressure on reef, (for example restricting the coastal settlement, development and tourism in coastal reef) and controlling pollution specially sewage wastage, may not only reduce coral declining but also reverse it and may let to coral reef more adaptable to changes relates to climate and acidification. The report shows that healthier reef in the Caribbean are those with large population of parrotfish in countries which protect these key fishes and sea urchins, banning fish trap and Spearfishing creating "resilient reefs".

To help combat ocean acidification, some laws are in place to reduce greenhouse gases such as carbon dioxide. The Clean Water Act puts pressure on state government agencies to monitor and limit runoff of pollutants that can cause ocean acidification. Stormwater surge preventions are also in place, as well as coastal buffers between agricultural land and the coastline. This act also ensures that delicate watershed ecosystems are intact, such as wetlands. The Clean Water Act is funded by the federal government, and is monitored by various watershed groups. Many land use laws aim to reduce CO_2 emissions by limiting deforestation. Deforestation causes erosion, which releases a large amount of carbon stored in the soil, which then flows into the ocean, contributing to ocean acidification. Incentives are used to reduce miles traveled by vehicles, which reduces the carbon emissions into the atmosphere, thereby reducing the amount of dissolved CO_2 in the ocean. State and federal governments also control coastal erosion, which releases stored carbon in the soil into the ocean, increasing ocean acidification. High-end satellite technology is increasingly being employed to monitor coral reef conditions.

Biosphere reserve, marine park, national monument and world heritage status can protect reefs. For example, Belize's barrier reef, Sian Ka'an, the Galapagos islands, Great Barrier Reef, Henderson Island, Palau and Papahānaumokuākea Marine National Monument are world heritage sites.

In Australia, the Great Barrier Reef is protected by the Great Barrier Reef Marine Park Authority, and is the subject of much legislation, including a biodiversity action plan. They have compiled a Coral Reef Resilience Action Plan. This detailed action plan consists of numerous adaptive management strategies, including reducing our carbon footprint, which would ultimately reduce the amount of ocean acidification in the oceans surrounding the Great Barrier Reef. An extensive public awareness plan is also in place to provide education on the "rainforests of the sea" and how people can reduce carbon emissions, thereby reducing ocean acidification.

Inhabitants of Ahus Island, Manus Province, Papua New Guinea, have followed a generations-old practice of restricting fishing in six areas of their reef lagoon. Their cultural traditions allow line fishing, but no net or spear fishing. The result is both the biomass and individual fish sizes are significantly larger than in places where fishing is unrestricted.

Restoration

Coral fragments growing on nontoxic concrete

Coral aquaculture, also known as coral farming or coral gardening, is showing promise as a potentially effective tool for restoring coral reefs, which have been declining around the world. The process bypasses the early growth stages of corals when they are most at risk of dying. Coral seeds are grown in nurseries, then replanted on the reef. Coral is farmed by coral farmers who live locally to the reefs and farm for reef conservation or for income.

Efforts to expand the size and number of coral reefs generally involve supplying substrate to allow more corals to find a home. Substrate materials include discarded vehicle tires, scuttled ships, subway cars, and formed concrete, such as reef balls. Reefs also grow unaided on marine structures such as oil rigs. In large restoration projects, propagated hermatypic coral on substrate can be secured with metal pins, superglue or milliput. Needle and thread can also attach A-hermatype coral to substrate.

A substrate for growing corals referred to as Biorock is produced by running low voltage electrical currents through seawater to crystallize dissolved minerals onto steel structures. The resultant white carbonate (aragonite) is the same mineral that makes up natural coral reefs. Corals rapidly colonize and grow at accelerated rates on these coated structures. The electrical currents also accelerate formation and growth of both chemical limestone rock and the skeletons of corals and other shell-bearing organisms. The vicinity of the anode and cathode provides a high-pH environment which inhibits the growth of competitive filamentous and fleshy algae. The increased growth rates fully depend on the accretion activity.

During accretion, the settled corals display an increased growth rate, size and density, but after the process is complete, growth rate and density return to levels comparable to natural growth, and are about the same size or slightly smaller.

One case study with coral reef restoration was conducted on the island of Oahu in Hawaii. The University of Hawaii has come up with a Coral Reef Assessment and Monitoring Program to help relocate and restore coral reefs in Hawaii. A boat channel on the island of Oahu to the Hawaii Institute of Marine Biology was overcrowded with coral reefs. Also, many areas of coral reef patches in the channel had been damaged from past dredging in the channel. Dredging covers the existing corals with sand, and their larvae cannot build and thrive on sand; they can only build on to existing reefs. Because of this, the University of Hawaii decided to relocate some of the coral reef to a different transplant site. They transplanted them with the help of the United States Army divers, to a relocation site relatively close to the channel. They observed very little, if any, damage occurred to any of the colonies while they were being transported, and no mortality of coral reefs has been observed on the new transplant site, but they will be continuing to monitor the new transplant site to see how potential environmental impacts (i.e. ocean acidification) will harm the overall reef mortality rate. While trying to attach the coral to the new transplant site, they found the coral placed on hard rock is growing considerably well, and coral was even growing on the wires that attached the transplant corals to the transplant site. This gives new hope to future research on coral reef transplant sites. As a result of this coral restoration project, no environmental effects were seen from the transplantation process, no recreational activities were decreased, and no scenic areas were affected by the project. This is a great example that coral transplantation and restoration can work and thrive under the right conditions, which means there may be hope for other damaged coral reefs.

Another possibility for coral restoration is gene therapy. Through infecting coral with genetically modified bacteria, it may be possible to grow corals that are more resistant to climate change and other threats.

Reefs in the Past

Ancient coral reefs

Throughout Earth history, from a few thousand years after hard skeletons were developed by marine organisms, there were almost always reefs. The times of maximum development were in the Middle Cambrian (513–501 Ma), Devonian (416–359 Ma) and Carboniferous (359–299 Ma), owing to order Rugosa extinct corals, and Late Cretaceous (100–66 Ma) and all Neogene (23 Ma–present), owing to order Scleractinia corals.

Not all reefs in the past were formed by corals: those in the Early Cambrian (542–513 Ma) resulted from calcareous algae and archaeocyathids (small animals with conical shape, probably related to sponges) and in the Late Cretaceous (100–66 Ma), when there also existed reefs formed by a group of bivalves called rudists; one of the valves formed the main conical structure and the other, much smaller valve acted as a cap.

Measurements of the oxygen isotopic composition of the aragonitic skeleton of coral reefs, such as Porites, can indicate changes in the sea surface temperature and sea surface salinity conditions of the ocean during the growth of the coral. This technique is often used by climate scientists to infer the paleoclimate of a region.

Demersal Zone

The demersal zone is the part of the sea or ocean (or deep lake) consisting of the part of the water column near to (and significantly affected by) the seabed and the benthos. The demersal zone is just above the benthic zone and forms a layer of the larger profundal zone.

Being just above the ocean floor, the demersal zone is variable in depth and can be part of the photic zone where light can penetrate and photosynthetic organisms grow, or the aphotic zone, which begins between depths of roughly 200 and 1,000 m (700 and 3,300 ft) and extends to the ocean depths, where no light penetrates.

Fish

The distinction between demersal species of fish and pelagic species is not always clear cut. The Atlantic cod (*Gadus morhua*) is a typical demersal fish, but can also be found in the open water column, and the Atlantic herring (*Clupea harengus*) is predominantly a pelagic species but forms large aggregations near the seabed when it spawns on banks of gravel.

Two types of fish inhabit the demersal zone, those that are heavier than water and rest on the seabed, and those that have neutral buoyancy and remain just above the substrate. In many species of fish, neutral buoyancy is maintained by a gas-filled swim bladder which can be expanded or contracted as the circumstances require. A disadvantage of this method is that adjustments need to be made constantly as the water pressure varies when the fish swims higher and lower in the water column. An alternative buoyancy aid is the use of lipids; these are less dense than water and squalene, commonly found in shark livers, has a specific gravity of just 0.86. In the velvet belly lanternshark (*Etmopterus spinax*), a benthopelagic species, 17% of the bodyweight is liver and 70% is lipids. Benthic rays and skates have smaller livers with lower concentrations of lipids; they are therefore denser than water and they do not swim continuously, intermittently resting on the seabed. Some fish have no buoyancy aids but use their pectoral fins which are so angled as to give lift as they swim. The disadvantage of this is that, if they stop swimming, the fish sink, and they cannot hover, or swim backwards.

Demersal fish have various feeding strategies; many feed on zooplankton or organisms or algae on the seabed; some of these feed on epifauna (invertebrates on top of the seafloor), while others specialise on infauna (invertebrates that burrow beneath the seafloor). Others are scavengers, eating the dead remains of plants or animals while still others are predators.

Invertebrates

Zooplankton are animals that drift with the current, but many have some limited means of locomotion and have some control over the depths at which they drift. They use gas-filled sacs or accumulations of substances with low densities to provide buoyancy, or they may have structures that slow down any passive descent. Where the adult, benthic organism is limited to life in a certain range of depths, their larvae need to optimise their chances of settling on a suitable substrate.

Cuttlefish are able to adjust their buoyancy using their cuttlebones, lightweight rigid structures with cavities filled with gas, which have a specific gravity of about 0.6. This enables them to swim at varying depths. Another invertebrate that feeds on the seabed and has swimming abilities is the nautilus, which stores gas in its chambers and adjusts its buoyancy by use of osmosis, pumping water in and out.

Critical Habitat

Critical habitat is a habitat area essential to the conservation of a listed species, though the area need not actually be occupied by the species at the time it is designated.

Designation Process

Critical habitat must be designated for all threatened species and endangered species under the Endangered Species Act, with certain specified exceptions. Designations of critical habitats must be based on:

- the best scientific information available

- in an open public process

- within specific timeframes.

Considerations

Before designating critical habitat, careful consideration must be given to the economic impacts, impacts on national security, and other relevant impacts of specifying any particular area as critical habitat. An area may be excluded from critical habitat if the benefits of exclusion outweigh the benefits of designation, unless excluding the area will result in the extinction of the species concerned.

If habitat land is nonfederal, there must be a federal connection for the ESA to be triggered; purely private actions are not covered. A federal agency with whom a landowner is dealing must ensure that its actions (which may include giving a loan, increasing irrigation flows, etc.) do not adversely modify these areas.

Desert

A desert is a barren area of landscape where little precipitation occurs and consequently living conditions are hostile for plant and animal life. The lack of vegetation exposes the unprotected surface of the ground to the processes of denudation. About one third of the land surface of the world is arid or semi-arid. This includes much of the polar regions where little precipitation occurs and which are sometimes called polar deserts or "cold deserts". Deserts can be classified by the amount of precipitation that falls, by the temperature that prevails, by the causes of desertification or by their geographical location.

Valle de la Luna ("Valley of the Moon") in the Atacama Desert of Chile, the world's driest hot desert

Sand dunes in the Rub' al Khali ("Empty quarter") of Saudi Arabia

Deserts are formed by weathering processes as large variations in temperature between day and night put strains on the rocks which consequently break in pieces. Although rain seldom occurs in deserts, there are occasional downpours that can result in flash floods. Rain falling on hot rocks can cause them to shatter and the resulting fragments and rubble strewn over the desert floor is further eroded by the wind. This picks up particles of sand and dust and wafts them aloft in sand or dust storms. Wind-blown sand grains striking any solid object in their path can abrade the surface. Rocks are smoothed down, and the wind sorts sand into uniform deposits. The grains end up as level sheets of sand or are piled high in billowing sand dunes. Other deserts are flat, stony plains where all the fine material has been blown away and the surface consists of a mosaic of smooth stones. These areas are known as desert pavements and little further erosion takes place. Other desert features include rock outcrops, exposed bedrock and clays once deposited by flowing water. Temporary lakes may form and salt pans may be left when waters evaporate. There may be underground sources of water in the form of springs and seepages from aquifers. Where these are found, oases can occur.

Plants and animals living in the desert need special adaptations to survive in the harsh environment. Plants tend to be tough and wiry with small or no leaves, water-resistant cuticles and often spines to deter herbivory. Some annual plants germinate, bloom and die in the course of a few weeks after rainfall while other long-lived plants survive for years and have deep root systems able to tap underground moisture. Animals need to keep cool and find enough food and water to survive. Many are nocturnal and stay in the shade or underground during the heat of the day. They tend to be efficient at conserving water, extracting most of their needs from their food and concentrating their urine. Some animals remain in a state of dormancy for long periods, ready to become active again when the rare rains fall. They then reproduce rapidly while conditions are favorable before returning to dormancy.

People have struggled to live in deserts and the surrounding semi-arid lands for millennia. Nomads have moved their flocks and herds to wherever grazing is available and oases have provided opportunities for a more settled way of life. The cultivation of semi-arid regions encourages erosion of soil and is one of the causes of increased desertification. Desert farming is possible with the aid of irrigation and the Imperial Valley in California provides an example of how previously barren land can be made productive by the import of water from an outside source. Many trade routes have been forged across deserts, especially across the Sahara Desert, and traditionally were used by caravans of camels carrying salt, gold, ivory and other goods. Large numbers of slaves were also taken northwards across the Sahara. Some mineral extraction also takes place in deserts and the uninterrupted sunlight gives potential for the capture of large quantities of solar energy.

Etymology

English *desert* and its Romance cognates all come from the ecclesiastical Latin *dēsertum* a participle of *dēserere*, "to abandon". The correlation between aridity and sparse population is complex and dynamic, varying by culture, era, and technologies; thus the use of the word *desert* can cause confusion. In English before the 20th century, *desert* was often used in the sense of "unpopulated area", without specific reference to aridity; but today the word is most often used in its climate-science sense (an area of low precipitation). Phrases such as "desert island" and "Great American Desert" in previous centuries did not necessarily imply sand or aridity; their focus was the sparse population.

Physical Geography

A desert is a region of land that is very dry because it receives low amounts of precipitation (usually in the form of rain but may be snow, mist or fog), often has little coverage by plants, and in which streams dry up unless they are supplied by water from outside the area. Deserts can also be described as areas where more water is lost by evapotranspiration than falls as precipitation. Deserts generally receive less than 250 mm (10 in) of precipitation each year. Semideserts are regions which receive between 250 and 500 mm (10 and 20 in) and when clad in grass, these are known as steppes.

Classification

Deserts have been defined and classified in a number of ways, generally combining total precipitation, number of days on which this falls, temperature, and humidity, and sometimes additional factors. For example, Phoenix, Arizona, receives less than 250 mm (9.8 in) of precipitation per year, and is immediately recognized as being located in a desert because of its aridity-adapted plants. The North Slope of Alaska's Brooks Range also receives less than 250 mm (9.8 in) of precipitation per year and is often classified as a cold desert. Other regions of the world have cold deserts, including areas of the Himalayas and other high-altitude areas in other parts of the world. Polar deserts cover much of the ice-free areas of the Arctic and Antarctic. A non-technical definition is that deserts are those parts of the Earth's surface that have insufficient vegetation cover to support a human population.

Potential evapotranspiration supplements the measurement of precipitation in providing a scientific measurement-based definition of a desert. The water budget of an area can be calculated using the formula $P - PE \pm S$, wherein P is precipitation, PE is potential evapotranspiration rates and S is amount of surface storage of water. Evapotranspiration is the combination of water loss through atmospheric evaporation and through the life processes of plants. Potential evapotranspiration, then, is the amount of water that *could* evaporate in any given region. As an example, Tucson, Arizona receives about 300 mm (12 in) of rain per year, however about 2,500 mm (98 in) of water could evaporate over the course of a year. In other words, about eight times more water could evaporate from the region than actually falls as rain. Rates of evapotranspiration in cold regions such as Alaska are much lower because of the lack of heat to aid in the evaporation process.

Deserts are sometimes classified as "hot" or "cold", "semiarid" or "coastal". The characteristics of hot deserts include high temperatures in summer; greater evaporation than precipitation usually exacerbated by high temperatures, strong winds and lack of cloud cover; considerable variation in the occurrence of precipitation, its intensity and distribution; and low humidity. Winter temperatures vary considerably between different deserts and are often related to the location of the desert on the continental landmass and the latitude. Daily variations in temperature can be as great as 22 °C (40 °F) or more, with heat loss by radiation at night being increased by the clear skies.

Cold desert: snow surface at Dome C Station, Antarctica

Cold deserts, sometimes known as temperate deserts, occur at higher latitudes than hot deserts, and the aridity is caused by the dryness of the air. Some cold deserts are far from the ocean and others are separated by mountain ranges from the sea and in both cases there is insufficient moisture in the air to cause much precipitation. The largest of these deserts are found in Central Asia. Others occur on the eastern side of the Rocky Mountains, the eastern side of the southern Andes and in southern Australia. Polar deserts are a particular class of cold desert. The air is very cold and carries little moisture so little precipitation occurs and what does fall, usually as snow, is carried along in the often strong wind and may form blizzards, drifts and dunes similar to those caused by dust and sand in other desert regions. In Antarctica, for example, the annual precipitation is about 50 mm (2 in) on the central plateau and some ten times that amount on some major peninsulas.

Based on precipitation alone, hyperarid deserts receive less than 25 mm (1 in) of rainfall a year; they have no annual seasonal cycle of precipitation and experience twelve-month periods with no rainfall at all. Arid deserts receive between 25 and 200 mm (1 and 8 in) in a year and semiarid deserts between 200 and 500 mm (8 and 20 in). However, such factors as the temperature, humidity, rate of evaporation and evapotranspiration, and the moisture storage capacity of the ground have a marked effect on the degree of aridity and the plant and animal life that can be sustained. Rain falling in the cold season may be more effective at promoting plant growth, and defining the boundaries of deserts and the semiarid regions that surround them on the grounds of precipitation alone is problematic.

A semi-arid desert or a steppe is a version of the arid desert with much more rainfall, vegetation and higher humidity. These regions feature a semi-arid climate and are less extreme than regular deserts. Like arid deserts, temperatures can vary greatly in semi deserts. They share some characteristics of a true desert and are usually located at the edge of deserts and continental dry areas. They usually receive precipitation from 250 mm (10 in) to 500 mm (20 in) but this can vary due to evapotranspiration and soil nutrition. Semi deserts can be found in the Tabernas Desert (and some of the Spanish Plateau), The Sahel, The Eurasian Steppe, most of Central Asia, the Western US, most of Northern Mexico, portions of South America (especially in Argentina) and the Australian Outback. They usually feature *BSh* (hot steppe) or *BSk* (temperate steppe) in the Köppen climate classification.

Coastal deserts are mostly found on the western edges of continental land masses in regions where cold currents approach the land or cold water upwellings rise from the ocean depths. The cool winds crossing this water pick up little moisture and the coastal regions have low temperatures and very low rainfall, the main precipitation being in the form of fog and dew. The range of temperatures on a daily and annual scale is relatively low, being 11 °C (20 °F) and 5 °C (9 °F) respectively in the Atacama Desert. Deserts of this type are often long and narrow and bounded to the east by mountain ranges. They occur in Namibia, Chile, southern California and Baja California. Other coastal deserts influenced by cold currents are found in Western Australia, the Arabian Peninsula and Horn of Africa, and the western fringes of the Sahara.

In 1961, Peveril Meigs divided desert regions on Earth into three categories according to the amount of precipitation they received. In this now widely accepted system, extremely arid lands have at least twelve consecutive months without precipitation, arid lands have less than 250 mm (10 in) of annual precipitation, and semiarid lands have a mean annual precipitation of between 250 and 500 mm (10–20 in). Both extremely arid and arid lands are considered to be deserts while semiarid lands are generally referred to as steppes when they are grasslands.

Deserts are also classified, according to their geographical location and dominant weather pattern, as trade wind, mid-latitude, rain shadow, coastal, monsoon, or polar deserts. Trade wind deserts occur either side of the horse latitudes at 30° to 35° North and South. These belts are associated with the subtropical anticyclone and the large-scale descent of dry air moving from high-altitudes toward the poles. The Sahara Desert is of this type. Mid-latitude deserts occur between 30° and 50° North and South. They are mostly in areas remote from the sea where most of the moisture has already precipitated from the prevailing winds. They include the Tengger and Sonoran Deserts. Monsoon deserts are similar. They occur in regions where large temperature differences occur between sea and land. Moist warm air rises over the land, deposits its water content and circulates back to sea. Further inland, areas receive very little precipitation. The Thar Desert near the India/ Pakistan border is of this type.

The Agasthiyamalai hills cut off Tirunelveli in India from the monsoons, creating a rainshadow region.

In some parts of the world, deserts are created by a rain shadow effect. Orographic lift occurs as air masses rise to pass over high ground. In the process they cool and lose much of their moisture by precipitation on the windward slope of the mountain range. When they descend on the leeward side, they warm and their capacity to hold moisture increases so an area with relatively little precipitation occurs. The Taklamakan Desert is an example, lying in the rain shadow of the Himalayas and receiving less than 38 mm (1.5 in) precipitation annually. Other areas are arid by virtue of being a very long way from the nearest available sources of moisture.

Montane deserts are arid places with a very high altitude; the most prominent example is found north of the Himalayas, in the Kunlun Mountains and the Tibetan Plateau. Many locations within this category have elevations exceeding 3,000 m (9,800 ft) and the thermal regime can be hemiboreal. These places owe their profound aridity (the average annual precipitation is often less than 40 mm or 1.5 in) to being very far from the nearest available sources of moisture and are often in the lee of mountain ranges. Montane deserts are normally cold, or may be scorchingly hot by day and very cold by night as is true of the northeastern slopes of Mount Kilimanjaro.

Polar deserts such as McMurdo Dry Valleys remain ice-free because of the dry katabatic winds that flow downhill from the surrounding mountains. Former desert areas presently in non-arid environments, such as the Sandhills in Nebraska, are known as paleodeserts. In the Köppen climate classification system, deserts are classed as *BWh* (hot desert) or *BWk* (temperate desert). In the Thornthwaite climate classification system, deserts would be classified as arid megathermal climates.

Weathering Processes

Deserts usually have a large diurnal and seasonal temperature range, with high daytime temperatures falling sharply at night. The diurnal range may be as much as 20 to 30 °C (36 to 54 °F) and the rock surface experiences even greater temperature differentials. During the day the sky is usually clear and most of the sun's radiation reaches the ground, but as soon as the sun sets, the desert cools quickly by radiating heat into space. In hot deserts, the temperature during daytime can exceed 45 °C (113 °F) in summer and plunge below freezing point at night during winter.

Exfoliation of weathering rocks in Texas

One square centimeter
(0.16 sq in) of windblown sand from the Gobi Desert.

Such large temperature variations have a destructive effect on the exposed rocky surfaces. The repeated fluctuations put a strain on exposed rock and the flanks of mountains crack and shatter. Fragmented strata slide down into the valleys where they continue to break into pieces due to the relentless sun by day and chill by night. Successive strata are exposed to further weathering. The relief of the internal pressure that has built up in rocks that have been underground for aeons can cause them to shatter. Exfoliation also occurs when the outer surfaces of rocks split off in flat flakes. This is believed to be caused by the stresses put on the rock by repeated expansions and contractions which induces fracturing parallel to the original surface. Chemical weathering processes probably play a more important role in deserts than was previously thought. The necessary moisture may be present in the form of dew or mist. Ground water may be drawn to the surface by evaporation and the formation of salt crystals may dislodge rock particles as sand or disintegrate rocks by exfoliation. Shallow caves are sometimes formed at the base of cliffs by this means.

As the desert mountains decay, large areas of shattered rock and rubble occur. The process continues and the end products are either dust or sand. Dust is formed from solidified clay or volcanic deposits whereas sand results from the fragmentation of harder granites, limestone and sandstone. There is a certain critical size (about 0.5 mm) below which further temperature-induced weathering of rocks does not occur and this provides a minimum size for sand grains.

As the mountains are eroded, more and more sand is created. At high wind speeds, sand grains are picked up off the surface and blown along, a process known as saltation. The whirling airborne grains act as a sand blasting mechanism which grinds away solid objects in its path as the kinetic energy of the wind is transferred to the ground. The sand eventually ends up deposited in level areas known as sand-fields or sand-seas, or piled up in dunes.

Dust Storms and saV Ndstorms

Dust storm about to engulf a military camp in Iraq, 2005

Sand and dust storms are natural events that occur in arid regions where the land is not protected by a covering of vegetation. Dust storms usually start in desert margins rather than the deserts themselves where the finer materials have already been blown away. As a steady wind begins to blow, fine particles lying on the exposed ground begin to vibrate. At greater wind speeds, some particles are lifted into the air stream. When they land, they strike other particles which may be jerked into the air in their turn, starting a chain reaction. Once ejected, these particles move in one of three possible ways, depending on their size, shape and density; suspension, saltation or creep. Suspension is only possible for particles less than 0.1 mm (0.004 in) in diameter. In a dust storm, these fine particles are lifted up and wafted aloft to heights of up to 6 km (3.7 mi). They reduce visibility and can remain in the atmosphere for days on end, conveyed by the trade winds for distances of up to 6,000 km (3,700 mi). Denser clouds of dust can be formed in stronger winds, moving across the land with a billowing leading edge. The sunlight can be obliterated and it may become as dark as night at ground level. In a study of a dust storm in China in 2001, it was estimated that 6.5 million tons of dust were involved, covering an area of 134,000,000 km² (52,000,000 sq mi). The mean particle size was 1.44 μm. A much smaller scale, short-lived phenomenon can occur in calm conditions when hot air near the ground rises quickly through a small pocket of cooler, low-pressure air above forming a whirling column of particles, a dust devil.

Sandstorms occur with much less frequency than dust storms. They are often preceded by severe dust storms and occur when the wind velocity increases to a point where it can lift heavier particles. These grains of sand, up to about 0.5 mm (0.020 in) in diameter are jerked into the air but soon fall back to earth, ejecting other particles in the process. Their weight prevents them from being airborne for long and most only travel a distance of a few meters (yards). The sand streams along above the surface of the ground like a fluid, often rising to heights of about 30 cm (12 in). In a really severe steady blow, 2 m (6 ft 7 in) is about as high as the sand stream can rise as the largest sand grains do not become airborne at all. They are transported by creep, being rolled along the desert floor or performing short jumps.

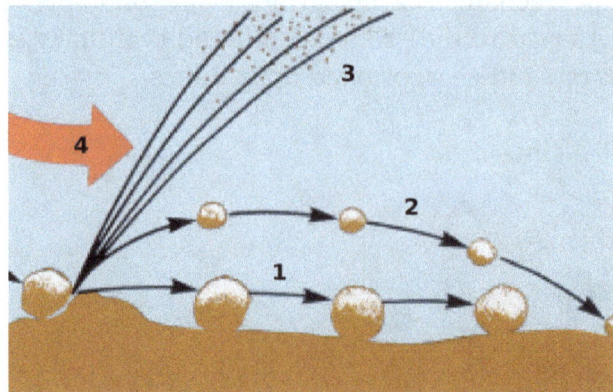

Wind-blown particles: 1=Creep 2=Saltation 3=Suspension 4=Wind current

During a sandstorm, the wind-blown sand particles become electrically charged. Such electric fields, which range in size up to 80 kV/m, can produce sparks and cause interference with telecommunications equipment. They are also unpleasant for humans and can cause headaches and nausea. The electric fields are caused by collision between airborne particles and by the impacts of saltating sand grains landing on the ground. The mechanism is little understood but the particles usually have a negative charge when their diameter is under 250 μm and a positive one when they are over 500 μm.

Major Deserts

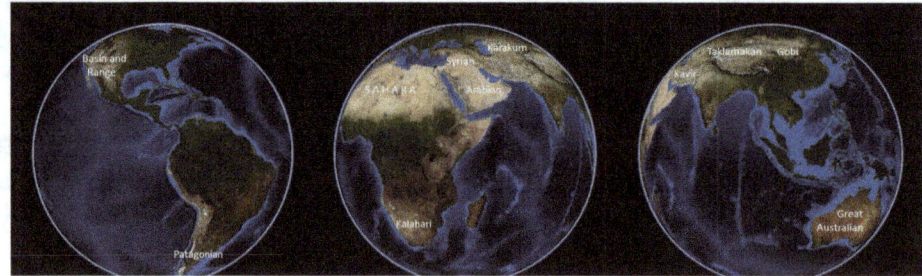

The world's largest non-polar deserts

Deserts take up about one third of the Earth's land surface. Bottomlands may be salt-covered flats. Eolian processes are major factors in shaping desert landscapes. Polar deserts (also seen as "cold deserts") have similar features, except the main form of precipitation is snow rather than rain. Antarctica is the world's largest cold desert (composed of about 98% thick continental ice sheet and 2% barren rock). Some of the barren rock is to be found in the so-called Dry Valleys of Antarctica that almost never get snow, which can have ice-encrusted saline lakes that suggest evaporation far greater than the rare snowfall due to the strong katabatic winds that even evaporate ice.

The ten largest deserts			
Rank	Desert	Area (km²)	Area (mi²)
1	Antarctic Desert (Antarctica)	14,200,000	5,500,000
2	Arctic Desert (Arctic)	13,900,000	5,400,000
3	Sahara Desert (Africa)	9,100,000	3,500,000

The ten largest deserts			
Rank	Desert	Area (km²)	Area (mi²)
4	Arabian Desert (Middle East)	2,600,000	1,000,000
5	Gobi Desert (Asia)	1,300,000	500,000
6	Patagonian Desert (South America)	670,000	260,000
7	Great Victoria Desert (Australia)	647,000	250,000
8	Kalahari Desert (Africa)	570,000	220,000
9	Great Basin Desert (North America)	490,000	190,000
10	Syrian Desert (Middle East)	490,000	190,000

Deserts, both hot and cold, play a part in moderating the Earth's temperature. This is because they reflect more of the incoming light and their albedo is higher than that of forests or the sea.

Features

Aerial view of Makhtesh Ramon, an erosion cirque of a type unique to the Negev.

Many people think of deserts as consisting of extensive areas of billowing sand dunes because that is the way they are often depicted on TV and in films, but deserts do not always look like this. Across the world, around 20% of desert is sand, varying from only 2% in North America to 30% in Australia and over 45% in Central Asia. Where sand does occur, it is usually in large quantities in the form of sand sheets or extensive areas of dunes.

A sand sheet is a near-level, firm expanse of partially consolidated particles in a layer that varies from a few centimeters to a few meters thick. The structure of the sheet consists of thin horizontal layers of coarse silt and very fine to medium grain sand, separated by layers of coarse sand and pea-gravel which are a single grain thick. These larger particles anchor the other particles in place and may also be packed together on the surface so as to form a miniature desert pavement. Small ripples form on the sand sheet when the wind exceeds 24 km/h (15 mph). They form perpendicular to the wind direction and gradually move across the surface as the wind continues to blow. The distance between their crests corresponds to the average length of jumps made by particles during saltation. The ripples are ephemeral and a change in wind direction causes them to reorganise.

Sand dunes are accumulations of windblown sand piled up in mounds or ridges. They form down-wind of copious sources of dry, loose sand and occur when topographic and climatic conditions cause airborne particles to settle. As the wind blows, saltation and creep take place on the wind-ward side of the dune and individual grains of sand move uphill. When they reach the crest, they cascade down the far side. The upwind slope typically has a gradient of 10° to 20° while the lee slope is around 32°, the angle at which loose dry sand will slip. As this wind-induced movement of sand grains takes place, the dune moves slowly across the surface of the ground. Dunes are sometimes solitary, but they are more often grouped together in dune fields. When these are extensive, they are known as sand seas or ergs.

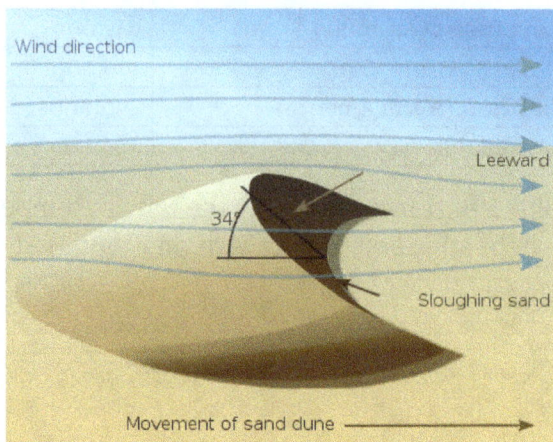

Diagram showing barchan dune formation, with the wind blowing from the left.

The shape of the dune depends on the characteristics of the prevailing wind. Barchan dunes are produced by strong winds blowing across a level surface, and are crescent-shaped with the concave side away from the wind. When there are two directions from which winds regularly blow, a series of long, linear dunes known as seif dunes may form. These also occur parallel to a strong wind that blows in one general direction. Transverse dunes run at a right angle to the prevailing wind direction. Star dunes are formed by variable winds, and have several ridges and slip faces radiating from a central point. They tend to grow vertically; they can reach a height of 500 m (1,600 ft), making them the tallest type of dune. Rounded mounds of sand without a slip face are the rare dome dunes, found on the upwind edges of sand seas.

Windswept desert pavement of small, smooth, closely packed stones in the Mojave desert.

A large part of the surface area of the world's deserts consists of flat, stone-covered plains dominated by wind erosion. In "eolian deflation", the wind continually removes fine-grained material, which becomes wind-blown sand. This exposes coarser-grained material, mainly pebbles with some larger stones or cobbles, leaving a desert pavement, an area of land overlaid by closely packed smooth stones forming a tessellated mosaic. Different theories exist as to how exactly the pavement is formed. It may be that after the sand and dust is blown away by the wind the stones jiggle themselves into place; alternatively, stones previously below ground may in some way work themselves to the surface. Very little further erosion takes place after the formation of a pavement, and the ground becomes stable. Evaporation brings moisture to the surface by capillary action and calcium salts may be precipitated, binding particles together to form a desert conglomerate. In time, bacteria that live on the surface of the stones accumulate a film of minerals and clay particles, forming a shiny brown coating known as desert varnish.

Other non-sandy deserts consist of exposed outcrops of bedrock, dry soils or aridisols, and a variety of landforms affected by flowing water, such as alluvial fans, sinks or playas, temporary or permanent lakes, and oases. A hamada is a type of desert landscape consisting of a high rocky plateau where the sand has been removed by aeolian processes. Other landforms include plains largely covered by gravels and angular boulders, from which the finer particles have been stripped by the wind. These are called "reg" in the western Sahara, "serir" in the eastern Sahara, "gibber plains" in Australia and "saï" in central Asia. The Tassili Plateau in Algeria is an impressive jumble of eroded sandstone outcrops, canyons, blocks, pinnacles, fissures, slabs and ravines. In some places the wind has carved holes or arches and in others it has created mushroom-like pillars narrower at the base than the top. In the Colorado Plateau it is water that has been the eroding force. Here the Colorado River has cut its way over the millennia through the high desert floor creating a canyon that is over a mile (6,000 feet or 1,800 meters) deep in places, exposing strata that are over two billion year old.

Water

Atacama, the world's driest non-polar desert, part of the Arid Diagonal of South America.

One of the driest places on Earth is the Atacama Desert. It is virtually devoid of life because it is blocked from receiving precipitation by the Andes mountains to the east and the Chilean Coast Range to the west. The cold Humboldt Current and the anticyclone of the Pacific are essential to keep the dry climate of the Atacama. The average precipitation in the Chilean region of Antofagasta is just 1 mm (0.039 in) per year. Some weather stations in the Atacama have never received rain. Evidence suggests that the Atacama may not have had any significant rainfall from 1570 to 1971. It is so arid that mountains that reach as high as 6,885 m (22,589 ft) are completely free of glaciers and, in the southern part from 25°S to 27°S, may have been glacier-free throughout the Quaternary, though permafrost extends down to an altitude of 4,400 m (14,400 ft) and is continuous above 5,600 m (18,400 ft). Nevertheless, there is some plant life in the Atacama, in the form of specialist plants that obtain moisture from dew and the fogs that blow in from the Pacific.

Flash flood in the Gobi

When rain falls in deserts, as it occasionally does, it is often with great violence. The desert surface is evidence of this with dry stream channels known as arroyos or wadis meandering across its surface. These can experience flash floods, becoming raging torrents with surprising rapidity after a storm that may be many kilometers away. Most deserts are in basins with no drainage to the sea but some are crossed by exotic rivers sourced in mountain ranges or other high rainfall areas beyond their borders. The River Nile, the Colorado River and the Yellow River do this, losing much of their water through evaporation as they pass through the desert and raising groundwater levels nearby. There may also be underground sources of water in deserts in the form of springs, aquifers, underground rivers or lakes. Where these lie close to the surface, wells can be dug and oases may form where plant and animal life can flourish. The Nubian Sandstone Aquifer System under the Sahara Desert is the largest known accumulation of fossil water. The Great Man-Made River is a scheme launched by Libya's Colonel Gadaffi to tap this aquifer and supply water to coastal cities. Kharga Oasis in Egypt is 150 km (93 mi) long and is the largest oasis in the Libyan Desert. A lake occupied this depression in ancient times and thick deposits of sandy-clay resulted. Wells are dug to extract water from the porous sandstone that lies underneath. Seepages may occur in the walls of canyons and pools may survive in deep shade near the dried up watercourse below.

Lakes may form in basins where there is sufficient precipitation or meltwater from glaciers above. They are usually shallow and saline, and wind blowing over their surface can cause stress, moving the water over nearby low-lying areas. When the lakes dry up, they leave a crust or hardpan behind. This area of deposited clay, silt or sand is known as a playa. The deserts of North America have more than one hundred playas, many of them relics of Lake Bonneville which covered parts of Utah, Nevada and Idaho during the last ice age when the climate was colder and wetter. These include the Great Salt Lake, Utah Lake, Sevier Lake and many dry lake beds. The smooth flat surfaces of playas have been used for attempted vehicle speed records at Black Rock Desert and Bonneville Speedway and the United States Air Force uses Rogers Dry Lake in the Mojave Desert as runways for aircraft and the space shuttle.

Biogeography

Flora

Xerophytes: Cardón cacti in the Baja California Desert, Cataviña region, Mexico.

Plants face severe challenges in arid environments. Problems they need to solve include how to obtain enough water, how to avoid being eaten and how to reproduce. Photosynthesis is the key to plant growth. It can only take place during the day as energy from the sun is required, but during the day, many deserts become very hot. Opening stomata to allow in the carbon dioxide necessary for the process causes evapotranspiration, and conservation of water is a top priority for desert vegetation. Some plants have resolved this problem by adopting crassulacean acid metabolism, allowing them to open their stomata during the night to allow CO_2 to enter, and close them during the day, or by using C4 carbon fixation.

A small tree in a sand hill, Polond Desert, Iran.

Many desert plants have reduced the size of their leaves or abandoned them altogether. Cacti are desert specialists and in most species the leaves have been dispensed with and the chlorophyll displaced into the trunks, the cellular structure of which has been modified to allow them to store water. When rain falls, the water is rapidly absorbed by the shallow roots and retained to allow them to survive until the next downpour, which may be months or years away. The giant saguaro cacti of the Sonoran Desert form "forests", providing shade for other plants and nesting places for desert birds. Saguaro grow slowly but may live for up to two hundred years. The surface of the trunk is folded like a concertina, allowing it to expand, and a large specimen can hold eight tons of water after a good downpour.

Cacti are present in both North and South America with a post-Gondwana origin. Other xerophytic plants have developed similar strategies by a process known as convergent evolution. They limit water loss by reducing the size and number of stomata, by having waxy coatings and hairy or tiny leaves. Some are deciduous, shedding their leaves in the driest season, and others curl their leaves up to reduce transpiration. Others store water in succulent leaves or stems or in fleshy tubers. Desert plants maximize water uptake by having shallow roots that spread widely, or by developing long taproots that reach down to deep rock strata for ground water. The saltbush in Australia has succulent leaves and secretes salt crystals, enabling it to live in saline areas. In common with cacti, many have developed spines to ward off browsing animals.

The camel thorn tree (*Acacia erioloba*) in the Namib Desert is nearly leafless in dry periods.

Some desert plants produce seed which lies dormant in the soil until sparked into growth by rainfall. When annuals, such plants grow with great rapidity and may flower and set seed within weeks, aiming to complete their development before the last vestige of water dries up. For perennial plants, reproduction is more likely to be successful if the seed germinates in a shaded position, but not so close to the parent plant as to be in competition with it. Some seed will not germinate until it has been blown about on the desert floor to scarify the seed coat. The seed of the mesquite tree, which grows in deserts in the Americas, is hard and fails to sprout even when planted carefully. When it has passed through the gut of a pronghorn it germinates readily, and the little pile of moist dung provides an excellent start to life well away from the parent tree. The stems and leaves of some plants lower the surface velocity of sand-carrying winds and protect the ground from erosion. Even small fungi and microscopic plant organisms found on the soil surface (so-called *cryptobiotic soil*) can be a vital link in preventing erosion and providing support for other living organisms. Some plants, including the Plantago Lanceolata, have to reproduce via wind pollination due to living in the environment. Cold deserts often have high concentrations of salt in the soil. Grasses and low shrubs are the dominant vegetation here and the ground may be covered with lichens. Most shrubs have spiny leaves and shed them in the coldest part of the year.

Fauna

Animals adapted to live in deserts are called xerocoles. There is no evidence that body temperature of mammals and birds is adaptive to the different climates, either of great heat or cold. In fact, with a very few exceptions, their basal metabolic rate is determined by body size, irrespective of the climate in which they live. Many desert animals (and plants) show especially clear evolutionary adaptations for water conservation or heat tolerance and so are often studied in comparative physiology, ecophysiology, and evolutionary physiology. One well-studied example is the specializations of mammalian kidneys shown by desert-inhabiting species. Many examples of convergent evolution have been identified in desert organisms, including between cacti and Euphorbia, kangaroo rats and jerboas, *Phrynosoma* and *Moloch* lizards.

The cream-colored courser, *Cursorius cursor*, is a well-camouflaged desert resident with its dusty coloration, countershading, and disruptive head markings.

Deserts present a very challenging environment for animals. Not only do they require food and water but they also need to keep their body temperature at a tolerable level. In many ways birds are the most able to do this of the higher animals. They can move to areas of greater food availability as the desert blooms after local rainfall and can fly to faraway waterholes. In hot deserts, gliding birds can remove themselves from the over-heated desert floor by using thermals to soar in the cooler air at great heights. In order to conserve energy, other desert birds run rather than fly. The cream-colored courser flits gracefully across the ground on its long legs, stopping periodically to snatch up insects. Like other desert birds it is well-camouflaged by its coloring and can merge into the landscape when stationary. The sandgrouse is an expert at this and nests on the open desert floor dozens of kilometers (miles) away from the waterhole it needs to visit daily. Some small diurnal birds are found in very restricted localities where their plumage matches the color of the underlying surface. The desert lark takes frequent dust baths which ensures that it matches its environment.

Water and carbon dioxide are metabolic end products of oxidation of fats, proteins, and carbohydrates. Oxidising a gram of carbohydrate produces 0.60 grams of water; a gram of protein produces 0.41 grams of water; and a gram of fat produces 1.07 grams of water, making it possible for xerocoles to live with little or no access to drinking water. The kangaroo rat for example makes use of this water of metabolism and conserves water both by having a low basal metabolic rate and by remaining underground during the heat of the day, reducing loss of water through its skin and respiratory system when at rest. Herbivorous mammals obtain moisture from the plants they eat. Species such as the addax antelope, dik-dik, Grant's gazelle and oryx are so efficient at doing this that they apparently never need to drink. The camel is a superb example of a mammal adapted to desert life. It minimizes its water loss by producing concentrated urine and dry dung, and is able to lose 40% of its body weight through water loss without dying of dehydration. Carnivores can obtain much of their water needs from the body fluids of their prey. Many other hot desert animals are nocturnal, seeking out shade during the day or dwelling underground in burrows. At depths of more than 50 cm (20 in), these remain at between 30 to 32 °C (86 to 90 °F) regardless of the external temperature. Jerboas, desert rats, kangaroo rats and other small rodents emerge from their burrows at night and so do the foxes, coyotes, jackals and snakes that prey on them. Kangaroos keep cool by increasing their respiration rate, panting, sweating and moistening the skin of their forelegs with saliva. Mammals living in cold deserts have developed greater insulation through warmer body fur and insulating layers of fat beneath the skin. The arctic weasel has a metabolic rate that is two or three times as high as would be expected for an animal of its size. Birds have avoided the problem of losing heat through their feet by not attempting to maintain them at the same temperature as the rest of their bodies, a form of adaptive insulation. The emperor penguin has dense plumage, a downy under layer, an air insulation layer next the skin and various thermoregulatory strategies to maintain its body temperature in one of the harshest environments on Earth.

The desert iguana (*Dipsosaurus dorsalis*) is well-adapted to desert life.

Being ectotherms, reptiles are unable to live in cold deserts but are well-suited to hot ones. In the heat of the day in the Sahara, the temperature can rise to 50 °C (122 °F). Reptiles cannot survive at this temperature and lizards will be prostrated by heat at 45 °C (113 °F). They have few adaptations to desert life and are unable to cool themselves by sweating so they shelter during the heat of the day. In the first part of the night, as the ground radiates the heat absorbed during the day, they emerge and search for prey. Lizards and snakes are the most numerous in arid regions and certain snakes have developed a novel method of locomotion that enables them to move sidewards and navigate high sand-dunes. These include the horned viper of Africa and the sidewinder of North America, evolutionarily distinct but with similar behavioural patterns because of convergent evolution. Many desert reptiles are ambush predators and often bury themselves in the sand, waiting for prey to come within range.

Amphibians might seem unlikely desert-dwellers, because of their need to keep their skins moist and their dependence on water for reproductive purposes. In fact, the few species that are found in this habitat have made some remarkable adaptations. Most of them are fossorial, spending the hot dry months aestivating in deep burrows. While there they shed their skins a number of times and retain the remnants around them as a waterproof cocoon to retain moisture. In the Sonoran Desert,

Couch's spadefoot toad spends most of the year dormant in its burrow. Heavy rain is the trigger for emergence and the first male to find a suitable pool calls to attract others. Eggs are laid and the tadpoles grow rapidly as they must reach metamorphosis before the water evaporates. As the desert dries out, the adult toads rebury themselves. The juveniles stay on the surface for a while, feeding and growing, but soon dig themselves burrows. Few make it to adulthood. The water holding frog in Australia has a similar life cycle and may aestivate for as long as five years if no rain falls. The Desert rain frog of Namibia is nocturnal and survives because of the damp sea fogs that roll in from the Atlantic.

Tadpole shrimp survive dry periods as eggs, which rapidly hatch and develop after rain.

Invertebrates, particularly arthropods, have successfully made their homes in the desert. Flies, beetles, ants, termites, locusts, millipedes, scorpions and spiders have hard cuticles which are impervious to water and many of them lay their eggs underground and their young develop away from the temperature extremes at the surface. The Saharan silver ant (*Cataglyphis bombycina*) uses a heat shock protein in a novel way and forages in the open during brief forays in the heat of the day. The long-legged darkling beetle in Namibia stands on its front legs and raises its carapace to catch the morning mist as condensate, funnelling the water into its mouth. Some arthropods make use of the ephemeral pools that form after rain and complete their life cycle in a matter of days. The desert shrimp does this, appearing "miraculously" in new-formed puddles as the dormant eggs hatch. Others, such as brine shrimps, fairy shrimps and tadpole shrimps, are cryptobiotic and can lose up to 92% of their bodyweight, rehydrating as soon as it rains and their temporary pools reappear.

Human Relations

Humans have long made use of deserts as places to live, and more recently have started to exploit them for minerals and energy capture. Deserts play a significant role in human culture with an extensive literature.

History

People have been living in deserts for millennia. Many, such as the Bushmen in the Kalahari, the Aborigines in Australia and various tribes of North American Indians, were originally hunter-gatherers. They developed skills in the manufacture and use of weapons, animal tracking, finding water, foraging for edible plants and using the things they found in their natural environment to supply their everyday needs. Their self-sufficient skills and knowledge were passed down through the generations by word of mouth. Other cultures developed a nomadic way of life as herders of sheep, goats, cattle, camels, yaks, llamas or reindeer. They travelled over large areas with their herds, moving to new pastures as seasonal and erratic rainfall encouraged new plant growth. They took with them their tents made of cloth or skins draped over poles and their diet included milk, blood and sometimes meat.

Shepherd near Marrakech leading his flock to new pasture

Salt caravan travelling between Agadez and the Bilma salt mines

The desert nomads were also traders. The Sahara is a very large expanse of land stretching from the Atlantic rim to Egypt. Trade routes were developed linking the Sahel in the south with the fertile Mediterranean region to the north and large numbers of camels were used to carry valuable goods across the desert interior. The Tuareg were traders and the goods transported traditionally included slaves, ivory and gold going northwards and salt going southwards. Berbers with knowledge of the region were employed to guide the caravans between the various oases and wells. Several million slaves may have been taken northwards across the Sahara between the 8th and 18th centuries. Traditional means of overland transport declined with the advent of motor vehicles, shipping and air freight, but caravans still travel along routes between Agadez and Bilma and between Timbuktu and Taoudenni carrying salt from the interior to desert-edge communities.

Round the rims of deserts, where more precipitation occurred and conditions were more suitable, some groups took to cultivating crops. This may have happened when drought caused the death of herd animals, forcing herdsmen to turn to cultivation. With few inputs, they were at the mercy of the weather and may have lived at bare subsistence level. The land they cultivated reduced the area available to nomadic herders, causing disputes over land. The semi-arid fringes of the desert have fragile soils which are at risk of erosion when exposed, as happened in the American Dust Bowl in the 1930s. The grasses that held the soil in place were ploughed under, and a series of dry years caused crop failures, while enormous dust storms blew the topsoil away. Half a million Americans were forced to leave their land in this catastrophe.

Similar damage is being done today to the semi-arid areas that rim deserts and about twelve million hectares of land are being turned to desert each year. Desertification is caused by such factors as drought, climatic shifts, tillage for agriculture, overgrazing and deforestation. Vegetation plays a major role in determining the composition of the soil. In many environments, the rate of erosion and run off increases dramatically with reduced vegetation cover. Unprotected dry surfaces tend to be blown away by the wind or be washed away by flash floods, leaving infertile soil layers that bake in the sun and become unproductive hardpan. Although overgrazing has historically been considered to be a cause of desertification, there is some evidence that wild and domesticated animals actually improve fertility and vegetation cover, and that their removal encourages erosive processes.

Natural Resource Extraction

A mining plant near Jodhpur, India

Deserts contain substantial mineral resources, sometimes over their entire surface, giving them their characteristic colors. For example, the red of many sand deserts comes from laterite minerals. Geological processes in a desert climate can concentrate minerals into valuable deposits. Leaching by ground water can extract ore minerals and redeposit them, according to the water table, in concentrated form. Similarly, evaporation tends to concentrate minerals in desert lakes, creating dry lake beds or playas rich in minerals. Evaporation can concentrate minerals as a variety of evaporite deposits, including gypsum, sodium nitrate, sodium chloride and borates. Evaporites are found in the USA's Great Basin Desert, historically exploited by the "20-mule teams" pulling carts of borax from Death Valley to the nearest railway. A desert especially rich in mineral salts is the Atacama Desert, Chile, where sodium nitrate has been mined for explosives and fertilizer since around 1850. Other desert minerals are copper from Chile, Peru, and Iran, and iron and uranium in Australia. Many other metals, salts and commercially valuable types of rock such as pumice are extracted from deserts around the world.

Oil and gas form on the bottom of shallow seas when micro-organisms decompose under anoxic conditions and later become covered with sediment. Many deserts were at one time the sites of shallow seas and others have had underlying hydrocarbon deposits transported to them by the movement of tectonic plates. Some major oilfields such as Ghawar are found under the sands of Saudi Arabia. Geologists believe that other oil deposits were formed by aeolian processes in ancient deserts as may be the case with some of the major American oil fields.

Farming

Traditional desert farming systems have long been established in North Africa, irrigation being the key to success in an area where water stress is a limiting factor to growth. Techniques that can be used include drip irrigation, the use of organic residues or animal manures as fertilisers and other traditional agricultural management practises. Once fertility has been built up, further crop production preserves the soil from destruction by wind and other forms of erosion. It has been found that plant growth-promoting bacteria play a role in increasing the resistance of plants to stress conditions and these rhizobacterial suspensions could be inoculated into the soil in the vicinity of the plants. A study of these microbes found that desert farming hampers desertification by establishing islands of fertility allowing farmers to achieve increased yields despite the adverse environmental conditions. A field trial in the Sonoran Desert which exposed the roots of different species of tree to rhizobacteria and the nitrogen fixing bacterium *Azospirillum brasilense* with the aim of restoring degraded lands was only partially successful.

The Judean Desert was farmed in the 7th century BC during the Iron Age to supply food for desert forts. Native Americans in the south western United States became agriculturalists around 600 AD when seeds and technologies became available from Mexico. They used terracing techniques and grew gardens beside seeps, in moist areas at the foot of dunes, near streams providing flood irrigation and in areas irrigated by extensive specially built canals. The Hohokam tribe constructed over 500 miles (800 km) of large canals and maintained them for centuries, an impressive feat of engineering. They grew maize, beans, squash and peppers.

Mosaic of fields in Imperial Valley

A modern example of desert farming is the Imperial Valley in California, which has high temperatures and average rainfall of just 3 in (76 mm) per year. The economy is heavily based on agriculture and the land is irrigated through a network of canals and pipelines sourced entirely from the Colorado River via the All-American Canal. The soil is deep and fertile, being part of the river's flood plains, and what would otherwise have been desert has been transformed into one of the most productive farming regions in California. Other water from the river is piped to urban communities but all this has been at the expense of the river, which below the extraction sites no longer has any above-ground flow during most of the year. Another problem of growing crops in this way is the build-up of salinity in the soil caused by evaporation of river water. The greening of the desert remains an aspiration and was at one time viewed as a future means for increasing food production for the world's growing population. This prospect has proved false as it disregarded the environmental damage caused elsewhere by the diversion of water for desert project irrigation.

Solar Energy Capture

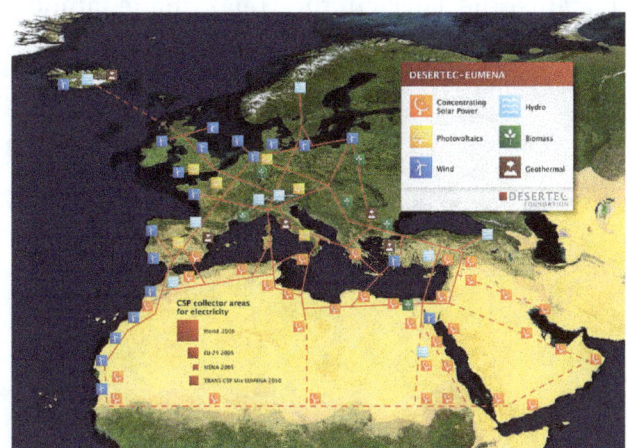

Desertec proposes using the Saharan and Arabian deserts to produce solar energy to power Europe and the Middle East.

Deserts are increasingly seen as sources for solar energy, partly due to low amounts of cloud cover. Many successful solar power plants have been built in the Mojave Desert. These plants have a combined capacity of 354 megawatts (MW) making them the largest solar power installation in the world. Large swaths of this desert are covered in mirrors, including nine fields of solar collectors. The Mojave Solar Park is currently under construction and will produce 280MW when completed.

The potential for generating solar energy from the Sahara Desert is huge, the highest found on the globe. Professor David Faiman of Ben-Gurion University has stated that the technology now exists to supply all of the world's electricity needs from 10% of the Sahara Desert. Desertec Industrial Initiative is a consortium seeking $560 billion to invest in North African solar and wind installations over the next forty years to supply electricity to Europe via cable lines running under the Mediterranean Sea. European interest in the Sahara Desert stems from its two aspects: the almost continual daytime sunshine and plenty of unused land. The Sahara receives more sunshine per acre than any part of Europe. The Sahara Desert also has the empty space totalling hundreds of square miles required to house fields of mirrors for solar plants.

The Negev Desert, Israel, and the surrounding area, including the Arava Valley, receive plenty of sunshine and are generally not arable. This has resulted in the construction of many solar plants. David Faiman has proposed that "giant" solar plants in the Negev could supply all of Israel's needs for electricity.

Warfare

The Arabs were probably the first organized force to conduct successful battles in the desert. By knowing back routes and the locations of oases and by utilizing camels, Muslim Arab forces were able to successfully overcome both Roman and Persian forces in the period 600 to 700 AD during the expansion of the Islamic caliphate.

Many centuries later, both world wars saw fighting in the desert. In the First World War, the Ottoman Turks were engaged with the British regular army in a campaign that spanned the Arabian peninsula. The Turks were defeated by the British, who had the backing of irregular Arab forces that were seeking to revolt against the Turks in the Hejaz, made famous in T. E. Lawrence's book *Seven Pillars of Wisdom.*

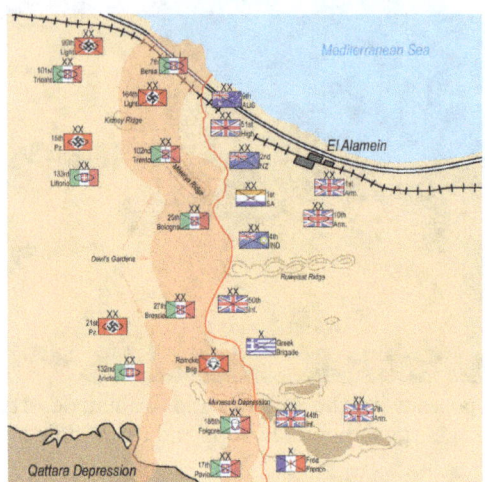

Deployment of forces on the eve of the Second Battle of El Alamein in 1942.

In the Second World War, the Western Desert Campaign began in Italian Libya. Warfare in the desert offered great scope for tacticians to use the large open spaces without the distractions of casualties among civilian populations. Tanks and armoured vehicles were able to travel large distances unimpeded and land mines were laid in large numbers. However the size and harshness of the terrain meant that all supplies needed to be brought in from great distances. The victors in a battle would advance and their supply chain would necessarily become longer, while the defeated army could retreat, regroup and resupply. For these reasons, the front line moved back and forth through hundreds of kilometers as each side lost and regained momentum. Its most easterly point was at El Alamein in Egypt, where the Allies decisively defeated the Axis forces in 1942.

In Culture

Marco Polo arriving in a desert land with camels. 14th century miniature from *Il milione.*

The desert is generally thought of as a barren and empty landscape. It has been portrayed by writers, film-makers, philosophers, artists and critics as a place of extremes, a metaphor for anything from death, war or religion to the primitive past or the desolate future.

There is an extensive literature on the subject of deserts. An early historical account is that of

Marco Polo (c. 1254–1324), who travelled through Central Asia to China, crossing a number of deserts in his twenty four year trek. Some accounts give vivid descriptions of desert conditions, though often accounts of journeys across deserts are interwoven with reflection, as is the case in Charles Montagu Doughty's major work, *Travels in Arabia Deserta* (1888). Antoine de Saint-Exupéry described both his flying and the desert in *Wind, Sand and Stars* and Gertrude Bell travelled extensively in the Arabian desert in the early part of the 20th century, becoming an expert on the subject, writing books and advising the British government on dealing with the Arabs. Another woman explorer was Freya Stark who travelled alone in the Middle East, visiting Turkey, Arabia, Yemen, Syria, Persia and Afghanistan, writing over twenty books on her experiences. The German naturalist Uwe George spent several years living in deserts, recording his experiences and research in his book, *In the Deserts of this Earth*.

The American poet Robert Frost expressed his bleak thoughts in his poem, *Desert Places*, which ends with the stanza "They cannot scare me with their empty spaces / Between stars - on stars where no human race is. / I have it in me so much nearer home / To scare myself with my own desert places."

Deserts on Other Planets

View of the Martian desert seen by the probe *Spirit* in 2004

Mars is the only planet in the Solar System on which deserts have been identified. Despite its low surface atmospheric pressure (only 1/100 of that of the Earth), the patterns of atmospheric circulation on Mars have formed a sea of circumpolar sand more than 5 million km² (1.9 million sq mi) in area, much larger than deserts on Earth. The Martian deserts principally consist of dunes in the form of half-moons in flat areas near the permanent polar ice caps in the north of the planet. The smaller dune fields occupy the bottom of many of the craters situated in the Martian polar regions. Examination of the surface of rocks by laser beamed from the Mars Exploration Rover have shown a surface film that resembles the desert varnish found on Earth although it might just be surface dust. The surface of Titan, a moon of Saturn, also has a desert-like surface with dune seas.

References

- Fratantoni, D.; Richardson P. (2006). "The Evolution and Demise of North Brazil Current Rings". Journal of Physical Oceanography. 36 (7): 1241–1249. Bibcode:2006JPO....36.1241F. doi:10.1175/JPO2907.1

- Spalding, Mark, Corinna Ravilious, and Edmund Green (2001). World Atlas of Coral Reefs. Berkeley, CA: University of California Press and UNEP/WCMC ISBN 0520232550

- Roach, John (November 7, 2001). "Rich Coral Reefs in Nutrient-Poor Water: Paradox Explained?". National Geographic News. Retrieved April 5, 2011

- Spalding MD, Grenfell AM (1997). "New estimates of global and regional coral reef areas". Coral Reefs. 16 (4): 225–230. doi:10.1007/s003380050078

- Gregg, M. (1989). "Scaling turbulent dissipation in the thermocline". Journal of Geophysical Research. 9686–9698. 94: 9686. Bibcode:1989JGR....94.9686G. doi:10.1029/JC094iC07p09686

- Moyle, Peter B.; Joseph J. Cech (2004). Fishes : an introduction to ichthyology (Fifth ed.). Upper Saddle River, N.J.: Pearson/Prentice Hall. p. 556. ISBN 978-0-13-100847-2

- Tobin, Barry (2003) [1998]. "How the Great Barrier Reef was formed". Australian Institute of Marine Science. Retrieved November 22, 2006

- Chappell, John (17 July 1980). "Coral morphology, diversity and reef growth". Nature. 286 (5770): 249–252. Bibcode:1980Natur.286..249C. doi:10.1038/286249a0

- Smalley, I. J.; Vita-Finzi, C. (1968). "The formation of fine particles in sandy deserts and the nature of 'desert' loess". Journal of Sedimentary Petrology. 38 (3): 766–774. doi:10.1306/74d71a69-2b21-11d7-8648000102c1865d

- Crossland CJ (1983) "Dissolved nutrients in coral reef waters In DJ Barnes (Ed) Perspectives on coral reefs, pages 56–68, Australian Institute of Marine Science. ISBN 9780642895851

- McClellan, Kate; Bruno, John (2008). "Coral degradation through destructive fishing practices". Encyclopedia of Earth. Retrieved October 25, 2008

- Meylan, Anne (January 22, 1988). "Spongivory in Hawksbill Turtles: A Diet of Glass". Science. 239 (4838): 393–395. Bibcode:1988Sci...239..393M. PMID 17836872. doi:10.1126/science.239.4838.393

- Latham, J. (1964). "The electrification of snowstorms and sandstorms" (PDF). Quarterly Journal of the Royal Meteorological Society. 90 (383): 91–95. Bibcode:1964QJRMS..90...91L. doi:10.1002/qj.49709038310

- Osborne, Patrick L. (2000). Tropical Ecosystem and Ecological Concepts. Cambridge: Cambridge University Press. p. 464. ISBN 0-521-64523-9

- Pidwirny, Michael (2008). "CHAPTER 8: Introduction to the Hydrosphere (e) Cloud Formation Processes". Physical Geography. Archived from the original on 2008-12-20. Retrieved 2009-01-01

- Power, James H. (1989). "Sink or Swim: Growth Dynamics and Zooplankton Hydromechanics". The American Naturalist. 133 (5): 706–721. JSTOR 2462076. doi:10.1086/284946

- Moseley, Pope L. (1997). "Heat shock proteins and heat adaptation of the whole organism". Journal of Applied Physiology. 83 (5): 1413–1417

3

Fundamentals of Ecological Extinction

Extinction is the termination of a group of organisms or a species so that they cease to exist. Extinctions can be of two types, mass extinctions and isolated extinctions. Climate change and human intervention that results in habitat destruction are some of the main causes of extinction. The chapter on ecological extinction offers an insightful focus, keeping in mind the complex subject matter.

Extinction

In biology and ecology, extinction is the end of an organism or of a group of organisms (taxon), normally a species. The moment of extinction is generally considered to be the death of the last individual of the species, although the capacity to breed and recover may have been lost before this point. Because a species' potential range may be very large, determining this moment is difficult, and is usually done retrospectively. This difficulty leads to phenomena such as Lazarus taxa, where a species presumed extinct abruptly "reappears" (typically in the fossil record) after a period of apparent absence.

More than 99 percent of all species, amounting to over five billion species, that ever lived on Earth are estimated to be extinct. Estimates on the number of Earth's current species range from 10 million to 14 million, of which about 1.2 million have been documented and over 86 percent have not yet been described. More recently, in May 2016, scientists reported that 1 trillion species are estimated to be on Earth currently with only one-thousandth of one percent described.

Through evolution, species arise through the process of speciation—where new varieties of organisms arise and thrive when they are able to find and exploit an ecological niche—and species become extinct when they are no longer able to survive in changing conditions or against superior competition. The relationship between animals and their ecological niches has been firmly established. A typical species becomes extinct within 10 million years of its first appearance, although some species, called living fossils, survive with virtually no morphological change for hundreds of millions of years.

Mass extinctions are relatively rare events; however, isolated extinctions are quite common. Only recently have extinctions been recorded and scientists have become alarmed at the current high rate of extinctions. Most species that become extinct are never scientifically documented. Some scientists estimate that up to half of presently existing plant and animal species may become extinct by 2100.

A dagger symbol (†) next to a species name is often used to indicate its extinction.

Definition

A species is extinct when the last existing member dies. Extinction therefore becomes a certainty when there are no surviving individuals that can reproduce and create a new generation. A species may become functionally extinct when only a handful of individuals survive, which cannot reproduce due to poor health, age, sparse distribution over a large range, a lack of individuals of both sexes (in sexually reproducing species), or other reasons.

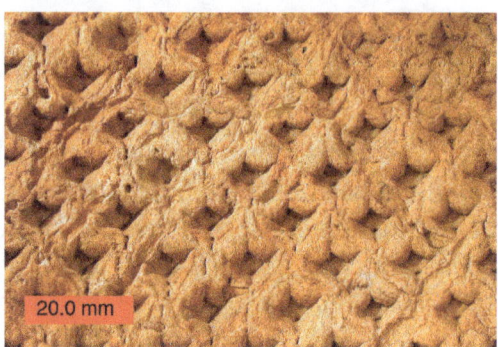

External mold of the extinct *Lepidodendron* from the Upper Carboniferous of Ohio.

Pinpointing the extinction (or pseudoextinction) of a species requires a clear definition of that species. If it is to be declared extinct, the species in question must be uniquely distinguishable from any ancestor or daughter species, and from any other closely related species. Extinction of a species (or replacement by a daughter species) plays a key role in the punctuated equilibrium hypothesis of Stephen Jay Gould and Niles Eldredge.

Skeleton of various extinct dinosaurs; some other dinosaur lineages still flourish in the form of birds.

In ecology, *extinction* is often used informally to refer to local extinction, in which a species ceases to exist in the chosen area of study, but may still exist elsewhere. This phenomenon is also known as extirpation. Local extinctions may be followed by a replacement of the species taken from other locations; wolf reintroduction is an example of this. Species which are not extinct are termed extant. Those that are extant but threatened by extinction are referred to as threatened or endangered species.

Currently an important aspect of extinction is human attempts to preserve critically endangered species. These are reflected by the creation of the conservation status "extinct in the wild" (EW). Species listed under this status by the International Union for Conservation of Nature (IUCN) are not known to have any living specimens in the wild, and are maintained only in zoos or other artificial environments. Some of these species are functionally extinct, as they are no longer part of their natural habitat and it is unlikely the species will ever be restored to the wild. When possible, modern zoological institutions try to maintain a viable population for species preservation and possible future reintroduction to the wild, through use of carefully planned breeding programs.

The dodo of Mauritius, shown here in a 1626 illustration by Roelant Savery, is an often-cited example of modern extinction.

The extinction of one species' wild population can have knock-on effects, causing further extinctions. These are also called "chains of extinction". This is especially common with extinction of keystone species.

Pseudoextinction

Extinction of a parent species where daughter species or subspecies are still extant is called pseudoextinction or phyletic extinction. Effectively, the old taxon vanishes, transformed (anagenesis) into a successor, or split into more than one (cladogenesis).

Pseudoextinction is difficult to demonstrate unless one has a strong chain of evidence linking a living species to members of a pre-existing species. For example, it is sometimes claimed that the extinct *Hyracotherium*, which was an early horse that shares a common ancestor with the modern horse, is pseudoextinct, rather than extinct, because there are several extant species of *Equus*, including zebra and donkey. However, as fossil species typically leave no genetic material behind, one cannot say whether *Hyracotherium* evolved into more modern horse species or merely evolved from a common ancestor with modern horses. Pseudoextinction is much easier to demonstrate for larger taxonomic groups.

Lazarus Taxa

The coelacanth, a fish related to lungfish and tetrapods, was considered to have been extinct since the end of the Cretaceous Period until 1938 when a specimen was found, off the Chalumna River (now Tyolomnqa) on the east coast of South Africa. Museum curator Marjorie Courtenay-Latimer discovered the fish among the catch of a local angler, Captain Hendrick Goosen, on December 23, 1938. A local chemistry professor, JLB Smith, confirmed the fish's importance with a famous cable: "MOST IMPORTANT PRESERVE SKELETON AND GILLS = FISH DESCRIBED".

Far more recent possible or presumed extinctions of species which may turn out still to exist include the thylacine, or Tasmanian tiger (*Thylacinus cynocephalus*), the last known example of which died in Hobart Zoo in Tasmania in 1936; the Japanese wolf (*Canis lupus hodophilax*), last sighted over 100 years ago; the ivory-billed woodpecker (*Campephilus principalis*), last sighted for certain in 1944; and the slender-billed curlew (*Numenius tenuirostris*), not seen since 2007.

Causes

The passenger pigeon, one of hundreds of species of extinct birds, was hunted to extinction over the course of a few decades.

As long as species have been evolving, species have been going extinct. It is estimated that over 99.9% of all species that ever lived are extinct. The average lifespan of a species is 1–10 million years, although this varies widely between taxa. There are a variety of causes that can contribute directly or indirectly to the extinction of a species or group of species. "Just as each species is unique", write Beverly and Stephen C. Stearns, "so is each extinction … the causes for each are varied—some subtle and complex, others obvious and simple". Most simply, any species that cannot survive and reproduce in its environment and cannot move to a new environment where it can do so, dies out and becomes extinct. Extinction of a species may come suddenly when an otherwise healthy species is wiped out completely, as when toxic pollution renders its entire habitat unliveable; or may occur gradually over thousands or millions of years, such as when a species gradually loses out in competition for food to better adapted competitors. Extinction may occur a long time after the events that set it in motion, a phenomenon known as extinction debt.

Assessing the relative importance of genetic factors compared to environmental ones as the causes of extinction has been compared to the debate on nature and nurture. The question of whether more extinctions in the fossil record have been caused by evolution or by catastrophe is a subject of discussion; Mark Newman, the author of *Modeling Extinction*, argues for a mathematical model that falls between the two positions. By contrast, conservation biology uses the extinction vortex model to classify extinctions by cause. When concerns about human extinction have been raised, for example in Sir Martin Rees' 2003 book *Our Final Hour*, those concerns lie with the effects of climate change or technological disaster.

Currently, environmental groups and some governments are concerned with the extinction of species caused by humanity, and they try to prevent further extinctions through a variety of conservation programs. Humans can cause extinction of a species through overharvesting, pollution, habitat destruction, introduction of invasive species (such as new predators and food competitors), overhunting, and other influences. Explosive, unsustainable human population growth is an essential cause of the extinction crisis. According to the International Union for Conservation of Nature (IUCN), 784 extinctions have been recorded since the year 1500, the arbitrary date selected to define "recent" extinctions, up to the year 2004; with many more likely to have gone unnoticed. Several species have also been listed as extinct since 2004.

Genetics and Demographic Phenomena

If adaptation increasing population fitness is slower than environmental degradation plus the accumulation of slightly deleterious mutations, then a population will go extinct. Smaller populations have fewer beneficial mutations entering the population each generation, slowing adaptation. It is also easier for slightly deleterious mutations to fix in small populations; the resulting positive feedback loop between small population size and low fitness can cause mutational meltdown.

Limited geographic range is the most important determinant of genus extinction at background rates but becomes increasingly irrelevant as mass extinction arises. Limited geographic range is a cause both of small population size and of greater vulnerability to local environmental catastrophes.

Extinction rates can be affected not just by population size, but by any factor that affects evolvability, including balancing selection, cryptic genetic variation, phenotypic plasticity, and robustness. A diverse or deep gene pool gives a population a higher chance in the short term of surviving an adverse change in conditions. Effects that cause or reward a loss in genetic diversity can increase the chances of extinction of a species. Population bottlenecks can dramatically reduce genetic diversity by severely limiting the number of reproducing individuals and make inbreeding more frequent.

Genetic Pollution

Scorched land resulting from slash-and-burn agriculture

Purebred wild species evolved to a specific ecology can be threatened with extinction through the process of genetic pollution—i.e., uncontrolled hybridization, introgression genetic swamping which leads to homogenization or out-competition from the introduced (or hybrid) species. Endemic populations can face such extinctions when new populations are imported or selectively bred by people, or when habitat modification brings previously isolated species into contact. Extinction is likeliest for rare species coming into contact with more abundant ones; interbreeding can swamp the rarer gene pool and create hybrids, depleting the purebred gene pool (for example, the endangered wild water buffalo is most threatened with extinction by genetic pollution from the abundant domestic water buffalo). Such extinctions are not always apparent from morphological (non-genetic) observations. Some degree of gene flow is a normal evolutionarily process, nevertheless, hybridization (with or without introgression) threatens rare species' existence.

The gene pool of a species or a population is the variety of genetic information in its living members. A large gene pool (extensive genetic diversity) is associated with robust populations that can survive bouts of intense selection. Meanwhile, low genetic diversity reduces the range of adaptions possible. Replacing native with alien genes narrows genetic diversity within the original population, thereby increasing the chance of extinction.

Habitat Degradation

Habitat degradation is currently the main anthropogenic cause of species extinctions. The main cause of habitat degradation worldwide is agriculture, with urban sprawl, logging, mining and some fishing practices close behind. The degradation of a species' habitat may alter the fitness landscape to such an extent that the species is no longer able to survive and becomes extinct. This may occur by direct effects, such as the environment becoming toxic, or indirectly, by limiting a species' ability to compete effectively for diminished resources or against new competitor species.

Habitat degradation through toxicity can kill off a species very rapidly, by killing all living members through contamination or sterilizing them. It can also occur over longer periods at lower toxicity levels by affecting life span, reproductive capacity, or competitiveness.

Habitat degradation can also take the form of a physical destruction of niche habitats. The widespread destruction of tropical rainforests and replacement with open pastureland is widely cited as an example of this; elimination of the dense forest eliminated the infrastructure needed by many species to survive. For example, a fern that depends on dense shade for protection from direct sunlight can no longer survive without forest to shelter it. Another example is the destruction of ocean floors by bottom trawling.

The golden toad was last seen on May 15, 1989. Decline in amphibian populations is ongoing worldwide.

Diminished resources or introduction of new competitor species also often accompany habitat degradation. Global warming has allowed some species to expand their range, bringing unwelcome competition to other species that previously occupied that area. Sometimes these new competitors are predators and directly affect prey species, while at other times they may merely outcompete vulnerable species for limited resources. Vital resources including water and food can also be limited during habitat degradation, leading to extinction.

Predation, Competition and Disease

In the natural course of events, species become extinct for a number of reasons, including but not limited to: extinction of a necessary host, prey or pollinator, inter-species competition, inability to deal with evolving diseases and changing environmental conditions (particularly sudden changes) which can act to introduce novel predators, or to remove prey. Recently in geological time, humans have become an additional cause of extinction (many people would say premature extinction) of some species, either as a new mega-predator or by transporting animals and plants from one part of the world to another. Such introductions have been occurring for thousands of years, sometimes intentionally (e.g. livestock released by sailors on islands as a future source of food) and sometimes accidentally (e.g. rats escaping from boats). In most cases, the introductions are unsuccessful, but when an invasive alien species does become established, the consequences can be catastrophic. Invasive alien species can affect native species directly by eating them, competing with them, and introducing pathogens or parasites that sicken or kill them; or indirectly by destroying or degrading their habitat. Human populations may themselves act as invasive predators. According to the "overkill hypothesis", the swift extinction of the megafauna in areas such as Australia (40,000 years before present), North and South America (12,000 years before present), Madagascar, Hawaii (300–1000 CE), and New Zealand (1300–1500 CE), resulted from the sudden introduction of human beings to environments full of animals that had never seen them before, and were therefore completely unadapted to their predation techniques.

Coextinction

The large Haast's eagle and moa from New Zealand

Coextinction refers to the loss of a species due to the extinction of another; for example, the extinction of parasitic insects following the loss of their hosts. Coextinction can also occur when a species loses its pollinator, or to predators in a food chain who lose their prey. "Species coextinction is a manifestation of the interconnectedness of organisms in complex ecosystems ... While coextinction may not be the most important cause of species extinctions, it is certainly an insidious one". Coextinction is especially common when a keystone species goes extinct. Models suggest that co-extinction is the most common form of biodiversity loss. There may be a cascade of coextinction across the trophic levels. Such effects are most severe in mutualistic and parasitic relationships. An example of coextinction is the Haast's eagle and the moa: the Haast's eagle was a predator that became extinct because its food source became extinct. The moa were several species of flightless birds that were a food source for the Haast's eagle.

Climate Change

Extinction as a result of climate change has been confirmed by fossil studies. Particularly, the extinction of amphibians during the Carboniferous Rainforest Collapse, 305 million years ago. A 2003 review across 14 biodiversity research centers predicted that, because of climate change, 15–37% of land species would be "committed to extinction" by 2050. The ecologically rich areas that would potentially suffer the heaviest losses include the Cape Floristic Region, and the Caribbean Basin. These areas might see a doubling of present carbon dioxide levels and rising temperatures that could eliminate 56,000 plant and 3,700 animal species.

Mass Extinctions

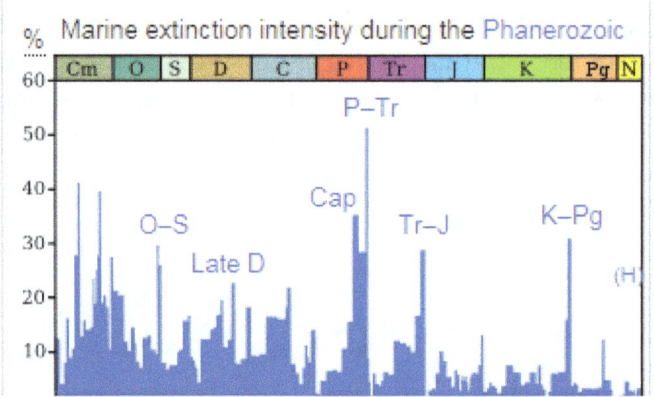

Millions of years ago. The blue graph shows the apparent *percentage* (not the absolute number) of marine animal genera becoming extinct during any given time interval. It does not represent all marine species, just those that are readily fossilized. The labels of the traditional "Big Five" extinction events and the more recently recognised End-Capitanian extinction event are clickable hyperlinks.

There have been at least five mass extinctions in the history of life on earth, and four in the last 350 million years in which many species have disappeared in a relatively short period of geological time. A massive eruptive event is considered to be one likely cause of the "Permian–Triassic extinction event" about 250 million years ago, which is estimated to have killed 90% of species then existing. There is also evidence to suggest that this event was preceded by another mass extinction, known as Olson's Extinction. The Cretaceous–Paleogene extinction event (K-Pg) occurred 66 million years ago, at the end of the Cretaceous period, and is best known for having wiped out non-avian dinosaurs, among many other species.

Modern Extinctions

According to a 1998 survey of 400 biologists conducted by New York's American Museum of Natural History, nearly 70% believed that the Earth is currently in the early stages of a human-caused mass extinction, known as the Holocene extinction. In that survey, the same proportion of respondents agreed with the prediction that up to 20% of all living populations could become extinct within 30 years (by 2028). A 2014 special edition of *Science* declared there is widespread consensus on the issue of human-driven mass species extinctions.

Biologist E. O. Wilson estimated in 2002 that if current rates of human destruction of the biosphere continue, one-half of all plant and animal species of life on earth will be extinct in 100 years. More significantly, the current rate of global species extinctions is estimated as 100 to 1000 times "background" rates (the average extinction rates in the evolutionary time scale of planet Earth), while future rates are likely 10,000 times higher. However, some groups are going extinct much faster. Biologists Paul R. Ehrlich and Stuart Pimm contend that human population growth is one of the main drivers of the modern extinction crisis.

History of Scientific Understanding

Dilophosaurus, one of the many extinct dinosaur genera. The cause of the Cretaceous–Paleogene extinction event is a subject of much debate amongst researchers.

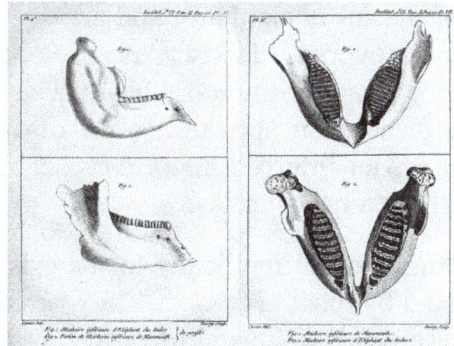

Georges Cuvier compared fossil mammoth jaws to those of living elephants, concluding that they were distinct from any known living species.

For much of history, the modern understanding of extinction as the end of a species was incompatible with the prevailing worldview. Through the 18th century, much of Western society adhered to the belief that the world was created by God and as such was complete and perfect. This concept reached its heyday in the 1700s with the peak popularity of a theological concept called the Great Chain of Being, in which all life on earth, from the tiniest microorganism to God, is linked in a continuous chain. The extinction of a species was impossible under this model, as it would create gaps or missing links in the chain and destroy the natural order. Thomas Jefferson was a firm supporter of the Great Chain of Being and an opponent of extinction, famously denying the extinction of the wooly mammoth on the grounds that nature never allows a race of animals to become extinct.

A series of fossils were discovered in the late 17th century that appeared unlike any living species. As a result, the scientific community embarked on a voyage of creative rationalization, seeking to understand what had happened to these species within a framework that did not account for total extinction. In October 1686, Robert Hooke presented an impression of a nautilus to the Royal Society that was more than two feet in diameter, and morphologically distinct from any known living species. Hooke theorized that this was simply because the species lived in the deep ocean and no one had discovered them yet. While he contended that it was possible a species could be "lost", he thought this highly unlikely. Similarly, in 1695, Thomas Molyneux published an account of enormous antlers found in Ireland that did not belong to any extant taxa in that area. Molyneux reasoned that they came from the North American moose and that the animal had once been common on the British Isles. Rather than suggest that this indicated the possibility of species going extinct, he argued that although organisms could become locally extinct, they could never be entirely lost and would continue to exist in some unknown region of the globe. Using the antlers as evidence for this position, Molyneux described how moose had continued to exist in North America even as they were lost to the British Isles. The antlers were later confirmed to be from the extinct Irish elk *Megaloceros*. Hooke and Molyneux's line of thinking was difficult to disprove. When parts of the world had not been thoroughly examined and charted, scientists could not rule out that animals found only in the fossil record were not simply "hiding" in unexplored regions of the Earth.

Georges Cuvier is credited with establishing the modern conception of extinction in a 1796 lecture to the French Institute, though he would spent most of his career trying to convince the wider scientific community of his theory. Cuvier was a well-regarded geologist, lauded for his ability to reconstruct the anatomy of an unknown species from a few fragments of bone. His primary evidence for extinction came from mammoth skulls found in the Paris basin. Cuvier recognized them as distinct from any known living species of elephant, and argued that it was highly unlikely such an enormous animal would go undiscovered. In 1812, Cuvier, along with Alexandre Bronigniart & Geoffroy Saint-Hilaire, mapped the strata of the Paris basin. They saw alternating saltwater and freshwater deposits, as well as patterns of the appearance and disappearance of fossils throughout the record. From these patterns, Cuvier inferred historic cycles of catastrophic flooding, extinction, and repopulation of the earth with new species.

Cuvier's fossil evidence showed that very different life forms existed in the past than those that exist today, a fact that was accepted by most scientists. The primary debate focused whether this turnover caused by extinction was gradual or abrupt in nature. Cuvier understood extinction to be the result of cataclysmic events that wipe out huge numbers of species, as opposed to the gradual decline of a species over time. His catastrophic view of the nature of extinction garnered him many opponents in the newly emerging school of uniformitarianism.

Jean-Baptist Lamarck, a gradualist and colleague of Cuvier, saw the fossils of different life forms as evidence of the mutable character of species. While Lamarck did not deny the possibility of extinction, he believed that it was exceptional and rare and that most of the change in species over time was due to gradual change. Unlike Cuvier, Lamarck was skeptical that catastrophic events of a scale large enough to cause total extinction were possible. In his geological history of the earth titled Hydrogeologie, Lamarck instead argued that the surface of the earth was shaped by gradual erosion and deposition by water, and that species changed over time in response to the changing environment.

Charles Lyell, a noted geologist and founder of uniformitarianism, believed that past processes should be understood using present day processes. Like Lamarck, Lyell acknowledged that extinction could occur, noting the total extinction of the dodo and the extirpation of indigenous horses to the British Isles. He similarly argued against mass extinctions, believing that any extinction must be a gradual process. Lyell also showed that Cuvier's original interpretation of the Parisian strata was incorrect. Instead of the catastrophic floods inferred by Cuvier, Lyell demonstrated that patterns of saltwater and freshwater deposits, like those seen in the Paris basin, could be formed by a slow rise and fall of sea levels.

The concept of extinction was integral to Charles Darwin's *On the Origin of Species*, with less fit lineages disappearing over time. For Darwin, extinction was a constant side effect of competition. Because of the wide reach of *On the Origin of Species*, it was widely accepted that extinction occurred gradually and evenly (a concept we now refer to as background extinction). It was not until 1982, when David Raup and Jack Sepkoski published their seminal paper on mass extinctions, that Cuvier was vindicated and catastrophic extinction was accepted as an important mechanism. The current understanding of extinction is a synthesis of the cataclysmic extinction events proposed by Cuvier, and the background extinction events proposed by Lyell and Darwin.

Human Attitudes and Interests

Extinction is an important research topic in the field of zoology, and biology in general, and has also become an area of concern outside the scientific community. A number of organizations, such as the Worldwide Fund for Nature, have been created with the goal of preserving species from extinction. Governments have attempted, through enacting laws, to avoid habitat destruction, agricultural over-harvesting, and pollution. While many human-caused extinctions have been accidental, humans have also engaged in the deliberate destruction of some species, such as dangerous viruses, and the total destruction of other problematic species has been suggested. Other species were deliberately driven to extinction, or nearly so, due to poaching or because they were "undesirable", or to push for other human agendas. One example was the near extinction of the American bison, which was nearly wiped out by mass hunts sanctioned by the United States government, to force the removal of Native Americans, many of whom relied on the bison for food.

Biologist Bruce Walsh of the University of Arizona states three reasons for scientific interest in the preservation of species; genetic resources, ecosystem stability, and ethics; and today the scientific community "stress[es] the importance" of maintaining biodiversity.

In modern times, commercial and industrial interests often have to contend with the effects of production on plant and animal life. However, some technologies with minimal, or no, proven harmful effects on *Homo sapiens* can be devastating to wildlife (for example, DDT). Biogeographer Jared Diamond notes that while big business may label environmental concerns as "exaggerated", and often cause "devastating damage", some corporations find it in their interest to adopt good conservation practices, and even engage in preservation efforts that surpass those taken by national parks.

Governments sometimes see the loss of native species as a loss to ecotourism, and can enact laws with severe punishment against the trade in native species in an effort to prevent extinction in the wild. Nature preserves are created by governments as a means to provide continuing habitats to species crowded by human expansion. The 1992 Convention on Biological Diversity has resulted in international Biodiversity Action Plan programmes, which attempt to provide comprehensive guidelines for government biodiversity conservation. Advocacy groups, such as The Wildlands Project and the Alliance for Zero Extinctions, work to educate the public and pressure governments into action.

People who live close to nature can be dependent on the survival of all the species in their environment, leaving them highly exposed to extinction risks. However, people prioritize day-to-day survival over species conservation; with human overpopulation in tropical developing countries, there has been enormous pressure on forests due to subsistence agriculture, including slash-and-burn agricultural techniques that can reduce endangered species's habitats.

Planned Extinction

Completed

- The smallpox virus is now extinct in the wild, although samples are retained in laboratory settings.

- The rinderpest virus, which infected domestic cattle, is now extinct in the wild.

Proposed

The poliovirus is now confined to small parts of the world due to extermination efforts.

Dracunculus medinensis, a parasitic worm which causes the disease dracunculiasis, is now close to eradication thanks to efforts led by the Carter Center.

Treponema pallidum pertenue, a bacterium which causes the disease yaws, is in the process of being eradicated.

Biologist Olivia Judson has advocated the deliberate extinction of certain disease-carrying mosquito species. In a September 25, 2003 *New York Times* article, she advocated "specicide" of thirty mosquito species by introducing a genetic element which can insert itself into another crucial gene, to create recessive "knockout genes". She says that the *Anopheles* mosquitoes (which spread malaria) and *Aedes* mosquitoes (which spread dengue fever, yellow fever, elephantiasis, and other diseases) represent only 30 species; eradicating these would save at least one million human lives per annum, at a cost of reducing the genetic diversity of the family Culicidae by only 1%. She further argues that since species become extinct "all the time" the disappearance of a few more will not destroy the ecosystem: "We're not left with a wasteland every time a species vanishes. Removing one species sometimes causes shifts in the populations of other species—but different need not mean worse." In addition, anti-malarial and mosquito control programs offer little realistic hope to the 300 million people in developing nations who will be infected with acute illnesses this year. Although trials are ongoing, she writes that if they fail: "We should consider the ultimate swatting."

Biologist E. O. Wilson has advocated the eradication of several species of mosquito, including malaria vector Anopheles gambiae. Wilson stated, "I'm talking about a very small number of species that have co-evolved with us and are preying on humans, so it would certainly be acceptable to remove them. I believe it's just common sense."

Cloning

Some, such as Harvard geneticist George M. Church, believe that ongoing technological advances will let us "bring back to life" an extinct species by cloning, using DNA from the remains of that species. Proposed targets for cloning include the mammoth, the thylacine, and the Pyrenean ibex. For this to succeed, enough individuals would have to be cloned, from the DNA of different individuals (in the case of sexually reproducing organisms) to create a viable population. Though bioethical and philosophical objections have been raised, the cloning of extinct creatures seems theoretically possible.

In 2003, scientists tried to clone the extinct Pyrenean ibex (*C. p. pyrenaica*). This attempt failed: of the 285 embryos reconstructed, 54 were transferred to 12 mountain goats and mountain goat-domestic goat hybrids, but only two survived the initial two months of gestation before they too died. In 2009, a second attempt was made to clone the Pyrenean ibex: one clone was born alive, but died seven minutes later, due to physical defects in the lungs.

Coextinction

Coextinction and cothreatened refer to the phenomena of the loss or decline of a host species resulting in the loss or endangerment of other species that depend on it, potentially leading to cascading effects across trophic levels. The term originated by the authors Stork and Lyal (1993) and was originally used to explain the extinction of parasitic insects following the loss of their specific hosts. The term is now used to describe the loss of any interacting species, including competition with their counterpart, and specialist herbivores with their food source. Coextinction is especially common when a keystone species goes extinct.

Causes

The most often cited example is that of the extinct passenger pigeon and its parasitic bird lice *Columbicola extinctus* and *Campanulotes defectus*. Recently, *C. extinctus* was rediscovered on the band-tailed pigeon, and *C. defectus* was found to be a likely case of misidentification of the existing *Campanulotes flavus*. However, even though the passenger pigeon lice story has a happy ending (i.e. rediscovery), coextinctions of other parasites, even on the passenger pigeon, may have occurred. Several louse species—such as *Rallicola extinctus*, a huia parasite—probably became extinct together with their hosts.

In recent studies, up to 50% of species have been said to go extinct in the next 50 years. This may be possible due to an example of coextinction being the loss of tropical butterfly species from Singapore attributing to the loss of their specific larval host plants. To see how possible future cases of coextinction would play out, researchers have made models to show probabilistic relationships between affiliate and host extinctions across co-evolved inter-specific systems. The subjects are pollinating Ficus Wasps and Ficus, primate parasites, (Pneumocystis Fungi, Nematode, and Lice) and their hosts, parasitic mites and lice and their avian hosts, butterflies and their larval host plants, and ant butterflies and their host ants. For all but the most host-specific affiliate groups (e.g., primate Pneumocystis fungi and primates), affiliate extinction levels may be modest at low levels of host extinction but can be expected to rise quickly as host extinctions increase to levels predicted in the near future. This curvilinear relationship between host and affiliate extinction levels may also explain, in part, why so few coextinction events have been documented to date (Koh *et al* 2004).

Recent investigations of coextinction risk among the rich Psyllid fauna Hemiptera: Psylloidea inhabiting acacias (Fabaceae-Mimosoideae: Acacia) in central eastern New South Wales, Australia provide information about coextinction. The results, derived from simple criteria to name host specialists among threatened acacias in Australia's central east, suggest that A. ausfeldii hosts one specialist psyllid species, Acizzia, and that A. gordonii hosts one specialist psyllid, Acizzia. Both psyllid species may be threatened at the same level of their host species with coextinction.

Interaction patterns can be used to anticipate the consequences of phylogenetic effects. By using a system of methodical observations, scientists can use the phylogenetic relationships of species to predict the number of interactions they exhibit in more than one-third of the networks, and the identity of the species with which they interact in about half of the networks. Consequentially, simulated extinction events tend to trigger coextinction cascades of related species. This results in a non-random pruning of the evolutionary tree.

In a 2004 paper in *Science*, ecologist Lian Pin Koh and colleagues discuss coextinction, stating

"Species coextinction is a manifestation of the interconnectedness of organisms in complex ecosystems. The loss of species through coextinction represents the loss of irreplaceable evolutionary and coevolutionary history. In view of the global extinction crisis, it is imperative that coextinction be the focus of future research to understand the intricate processes of species extinctions. While coextinction may not be the most important cause of species extinctions, it is certainly an insidious one." (Koh *et al.* 2004)

Koh *et al.* also define coendangered as taxon

"likely to go extinct if their currently endangered hosts [...] become extinct."

One example is the near extinction of the genus *Hibiscadelphus* as a consequence of the disappearance of several of the Hawaiian honeycreepers, its pollinators. There are several instances of predators and scavengers dying out following the disappearance of species which represented their source of food: for example, the coextinction of the Haast's eagle with the moa.

Coextinction may also occur on a local level: for example, the decline in the red ant *Myrmica sabuleti* in southern England, caused by habitat loss, resulted in the local extinction of the large blue butterfly, which is dependent on the ant as a host for the larvae. In this case the ant avoided local extinction, and the butterfly has been reintroduced.

Another example of species going through coextinction is rhinoceros stomach bot fly (Gyrostigma rhinocerontis) and of its host species the endangered black rhinoceros (Diceros bicornis) and the white rhinoceros (Ceratotherium simum). The female fly lays eggs behind the rhinoceros' ears or horns or neck and then the flies' larvae enter the animal's digestive tract and digs into the stomach lining and then are excreted out and the lifecycle restarts.

Consequences

Coextinction can mean loss of biodiversity and diversification. Coextinctions can influence not only parasite and mutualist diversification but also their hosts. Arguably, parasites facilitate host diversification through sexual selection. That loss of parasites can reduce host diversification rates. Coextinction can also result in loss of evolutionary history. The extinction of related hosts can lead to the extinction of related parasites. The loss of history is likely to be greater than the loss expected, were species to go extinct at random. Furthermore, if coextinctions are clustered, it is more likely that coextinction can produce non-random trait loss. Species that are at risk of coextinction are expected to be larger because rare hosts tend to be larger and larger hosts have larger parasites. They can also be expected to have lengthy generation times or higher tropic positions. Coextinction can extend beyond biodiversity and has direct and indirect consequences from the communities of lost species. One main consequence of coextinction that goes beyond biodiversity is mutualism, by loss of food production with a decline in threatened pollinators. Losses of parasites can have negative impacts on humans or the species. In rare hosts, losses of specialist parasites can predispose hosts to infection by emergent parasites. Furthermore, relating to the consequences of removing specialist parasites from rare hosts, is the problem of where the parasites will go once their host is extinct. If the parasites are dependent on only those species than there are parasite species that are at risk of extinction through co-endangerment. On the other hand, if they are able to find and switch onto alternative hosts, those hosts can turn out to be humans. Either way, the loss of parasites by co extinction or the acquiring of new parasites by alternative hosts, proves to be a major issue. Coextinction can go beyond the decreased biodiversity, it can range into various biomes and link various ecosystems.

A study conducted in New Caledonia has shown that extinction of a coral reef-associated fish species of average size would eventually result in the co-extinction of at least ten species of parasites.

Risks

The host Specificity and Life Cycle is a major factor in the risk of coextinction. Species of mutalists, parasites, and many free-living insects that have staged life cycles are more likely to be a victim of coextinction. This is due to the fact that these organisms may depend on multiple hosts throughout their lives in comparison to simple life cycled organisms. Also, if organisms are evolutionary flexible, then these organisms may escape extinction.

The area with that has the greatest affect of coextinction is the tropics. There is a continued disappearance in the habitat, human intervention, and a great loss in vital ecosystem services. This is threatening because the tropics contain 2/3 of the all known species but they aren't in a situation where they can be fully taken care of. Along with forest loss other risk factors include: coastal development, overexploitation of wildlife, and habitat conversion, that also affect human well-being.

In an effort to find a stop to coextinction, researchers have found that the first step would be to conserve the host species in which other species are dependent on. These hosts serve as major components for their habitat and need them to survive. In deciding what host to protect, it is important to choose one that can benefit an array of other dependent species.

Ecological Extinction

Ecological extinction is defined as "the reduction of a species to such low abundance that, although it is still present in the community, it no longer interacts significantly with other species."

Ecological extinction stands out because it is the interaction ecology of a species that is important for conservation work. They state that "unless the species interacts significantly with other species in the community (e.g. it is an important predator, competitor, symbiont, mutualist, or prey) its loss may result in little to no adjustment to the abundance and population structure of other species."

This view stems from the neutral model of communities that assumes there is little to no interaction within species unless otherwise proven.

Estes, Duggins, and Rathburn (1989) recognize two other distinct types of extinction:

- Global extinction is defined as "the ubiquitous disappearance of a species."

- Local extinction is characterized by "the disappearance of a species from part of its natural range."

Keystone Species

Paine first established the concept of a keystone species by studying the sea star.

Robert Paine (1969) first came up with the concept of a keystone species while studying the effects of the predatory sea star *Pisaster ochraceus*, on the abundance of the herbivorous gastropod, *Tegula funebralis*. This study took place in the rocky intertidal habitat off the coast of Washington;

Paine removed all *Pisaster* in 8m x 10m plots weekly while noting the response of *Tegula* for two years. He found that removing the top predator, in this case being *Pisaster*, reduced species number in the treatment plots. Paine defined the concept of a keystone species as a species that has a disproportionate effect on the community structure of an environment in relation to its total biomass. This keystone species effect forms the basis for the concept of ecological extinction.

Examples

Sea otters maintain the overall biodiversity of the kelp forest community.

Estes et al. (1978) evaluated the potential role of the sea otter as the keystone predator in nearshore kelp forests. They compared the Rat and Near islands in the Aleutian islands to test if "sea otter predation controls epibenthic invertebrate populations (specifically sea urchins), and in turn releases the vegetation association from intense grazing". Estes and his colleagues found that different size structures and densities of sea urchins were correlated with the presence of sea otter populations, and because they are the principal prey of this keystone predator, the sea otters were most likely the main determinants of the differences in sea urchin populations. With high sea otter densities the herbivory of sea urchins in these kelp forest was severely limited, and this made competition between algal species the main determinant in survival. However, when sea otters were absent, herbivory of the sea urchins was greatly intensified to the point of decimation of the kelp forest community. This loss of heterogeneity serves as a loss of habitat for both fish and eagle populations that depend on the richly productive kelp forest environment. Historical over harvesting of sea otter furs has severely restricted their once wide-ranging habitat, and only today are scientists starting to see the implications of these local extinctions. Conservation work needs to focus on finding the density threshold that render the sea otters an effective population. It must then continue and artificially repopulate the historical range of the sea otter in order to allow kelp forest communities to re-establish.

The California spiny lobster, or *Panulirus interruptus*, is another example of a keystone predator that has a distinct role in maintaining species diversity in its habitat. Robles (1987) demonstrated experimentally that the exclusion of spiny lobsters from the intertidal zone habitats led to the competitive dominance of mussels (*Mytilus edulis* and *M. californianus*). This results shows another example of how the ecological extinction of a keystone predator can reduce species diversity in an ecosystem. Unfortunately, the threshold of ecological extinction has long passed due to over fishing now that many local extinctions of the California spiny lobster are common.

Jackson et al. (2001) took a much needed historical perspective on the role of ecological extinction caused by overfishing of oysters in the Chesapeake Bay. Commercial oyster fishing had not affected the bay ecosystem until mechanical dredges for harvesting were utilized in the 1870s. The bay today is plagued by eutrophication due to algal blooms, and the resulting water is highly hypoxic. These algal blooms have competitively excluded any other species from surviving, including the rich diversity in faunal life that once flourished such as dolphins, manatees, river otters, sea turtles, alligators, sharks, and rays. This example highlights the top-down loss of diversity commercial fishing has on marine ecosystems by removing the keystone species of the environment.

Invasive Species

Christian found Argentine ants to disrupt large seed dispersal mutualisms.

Novaro et al. (2000) assessed the potential ecological extinction of guanacos (*Lama guanocoe*) and lesser rheas (*Pterocnemia pennata*) as a prey source for native omnivores and predators in the Argentine Patagonia. These native species are being replaced by introduced species such as the European rabbit, red deer, and domestic cattle; the cumulative damage from the increased herbivory by introduced species has also served to accelerate destruction of the already dwindling Argentine pampas and steppe habitats. This was the first study to take into account a large number of diverse predators, ranging from skunks to pumas, as well as conduct their survey in non-protected areas that represent the majority of southern South America. Novaro and his colleagues found that the entire assemblage of native carnivores relied primarily on introduced species as a prey base. They also suggested that the lesser rhea and guanaco had already passed their ecological effective density as a prey species, and thus were ecological extinct. It is possible that the niches of introduced species as herbivores too closely mirrored those of the natives, and thus competition was the primary cause of ecological extinction. The effect of introduction of new competitors, such as the red deer and rabbit, also served to alter the vegetation in the habitat, which could have further pronounced the intensity of competition. Guanacos and rheas have been classified as a low risk for global extinction, but this simplistic view of their demography doesn't take into account that they have already become functionally extinct in the Argentine Patagonia. Novaro and his colleagues suggest "this loss could have strong effects on plant-animal interactions, nutrient dynamics, and disturbance regimes ..." This is a prime example of how current conservation policy has already failed to protect the intended species because of its lack of a functionally sound definition for extinction.

Seed dispersal mechanisms play a fundamental role in the regeneration and continuation of community structure, and a recent study by Christian (2001) demonstrated a shift in the composition of the plant community in the South African shrublands following an invasion by the Argentine ant *(Linepithema humile)*. Ants disperse up to 30% of the flora in the shrublands and are vital to the survival of fynbos plants because they bury the large seeds away from the dangers of predation and fire damage. It is also crucial for seeds to be buried, because nearly all seed germination takes place in the first season after a fire. Argentine ants, a recent invader, do not disperse even small seeds. Christian tested whether the invasion of the Argentine ant differentially effected small and large-seeded fauna. He found that post-fire recruitment of large-seeded flora was reduced disproportionately for large seeds in sites already invaded by Argentine ants. These initial low large-seed density recruitments will eventually lead the domination of small-seeded fauna in invaded habitats. The consequences of this change in community structure highlight the struggle for dispersal of large-seeded flora that have potential reverberations around the world because ants are major ecological seed dispersers throughout the globe.

Modeling Ecological Extinction

Below a certain density threshold, flying foxes are no longer effective seed dispersers.

The McConkey and Drake (2006) study is unique because it was one of the first attempts to model a density-dependent threshold relationship that described ecological extinction. They studied a seed dispersal interaction between flying foxes and trees with large seeds on the tropical Pacific Islands. Insular flying foxes *(Pteropus tonganus)*, are considered to be keystone species because they are the only seed dispersers that can carry large seeds long distances. The host-pathogen model by Janzen and Connell suggests that survivorship of seeds in the tropics greatly increases the further away from the parent tree it lands, and that trees require this dispersal in order to avoid extinction. In the pathogen latent environment of the tropics, seed dispersal only becomes more paramount to species survival. As hypothesized, McConkey and Drake found a threshold relationship between the Flying Fox Index (FFI) and the median proportion of seeds carried over five meters. Below the threshold of abundance seed dispersal was insignificant and independent of flying fox abundance; however, above the threshold, dispersal positively correlated with increased flying fox abundance (as measured by the FFI). Although they did not directly prove the cause for this relationship, McConkey and Drake proposed a behavioral mechanism. Flying foxes are known to be territorial, and in the absence of competition a flying fox will eat within one tree, effectively dropping the seeds right below it. Alternatively, if there is a high density of flying foxes feeding at one time (abundance above the threshold density) then aggressive behavior, such as stealing fruit from another individual's territory, will lead to longer average seed dispersal. In this way the seed dispersing flying fox has a disproportional effect on the overall community structure in comparison to their relative biomass. Modeling the effect of ecological extinction on communities is the first step to applying this framework into conservation work.

While ecologists are just starting to get a grapple on the significant interactions within an ecosystem, they must continue to find an effective density threshold that can maintain the level of equilibrium species diversity. Only with this knowledge of where and to what extent a specific species interacts with its environment will the proper and most efficient levels of conservation work take place. This work is especially important on the limited ecosystems of islands, where there are less likely to be replacement species for specific niches. With species diversity and available habitat decreasing rapidly worldwide, identifying the systems that are most crucial to the ecosystem will be the crux of conservation work.

Climate Change

Climate change has produced numerous shifts in the distributions and abundances of species. Thomas et al. (2004) went on to assess the extinction risk due to these shifts over a broad range of global habitats. Their predictive model using midline estimates for climate warming over the next 50 years suggests that 15-37% of species will be "committed to extinction" be 2050. Although the average global temperature has risen .6°C, individual populations and habitats will only respond to their local changes in climate. Root et al. (2002) suggests that local changes in climate may account for density changes in regions, shifts in phenology (timing) of events, changes in morphology (biology) (such as body size), and shifts in genetic frequencies. They found that there have been an average phenological shift of 5.1 days earlier in the spring for a broad range of over a thousand compiled studies. This shift was also, as predicted, more pronounced in the upper latitudes that have concurrently had the largest shift in local average temperatures.

While the loss of habitat, loss of pollinator mutualisms, and the effect of introduced species all have distinct pressures on native populations, these effects must be looked underneath a synergistic and not an independent framework. Climate change has the potential to exacerbate all of these processes. Nehring (1999) found a total of 16 non-indigenous thermophilic phytoplankton established in habitats northwards of their normal range in the North Sea. He likened these changes in range of more southerly phytoplankton to climatic shifts in ocean temperature. All of these effects have additive effects to the stress on populations within an environment, and with the additionally fragile and more complete definition of ecological extinction must be taken into account into preventative conservation measures.

Implications for Conservation Policy

Conservation policy has historically lagged behind current science all over the world, but at this critical juncture politicians must make the effort to catch up before massive extinctions occur on our planet.

For example, the pinnacle of American conservation policy, the Endangered Species Act of 1973, fails to acknowledge any benefit for protecting highly interactive species that may help maintain overall species diversity. Policy must first assess whether the species in question is considered highly interactive by asking the questions "does the absence or loss of this species, either directly or indirectly, incur a loss of overall diversity, effect the reproduction or recruitment of other species, lead to a change in habitat structure, lead to a change in productivity or nutrient dynamics between ecosystems, change important ecological processes, or reduce the resilience of the ecosystem to disturbances?". After these multitudes of questions are addressed to define an interactive species, an ecologically effective density threshold must be estimated in order to maintain this interaction ecology. This process holds many of the same variables contained within viable population estimates, and thus should not be difficult to incorporate into policy. To avoid mass extinction on a global scale unlike anyone has seen before, scientists must understand all of the mechanisms driving the process. It is now that the governments of the world must act in order to prevent this catastrophe of the loss of biodiversity from progressing further and wasting all of the time and money spent on previous conservation efforts.

Local Extinction

Local extinction, or extirpation, is the condition of a species (or other taxon) that ceases to exist in the chosen geographic area of study, though it still exists elsewhere. Local extinctions are contrasted with global extinctions.

Local extinctions may be followed by a replacement of the species taken from other locations; wolf reintroduction is an example of this.

Conservation

Local extinctions mark a change in the ecology of an area.

The area of study chosen may reflect a natural subpopula, political boundaries, or both. The Cetacean Specialist Group of the IUCN has assessed the threat of a local extinction of the Black Sea stock of Harbour Porpoise (*Phocoena phocoena*) that touches six different countries. COSEWIC, by contrast, investigates wildlife only in Canada, so assesses only the risk of a Canadian local extinction even for species that cross into the United States or other countries. Other subpopulations may be naturally divided by political or country boundaries.

Many crocodilian species have experienced localized extinction, particularly the saltwater crocodile (*Crocodylus porosus*), which has been extirpated from Vietnam, Thailand, Java, and many other areas.

Often a subpopulation of a species will also be a subspecies. For example, the recent disappearance of the Black Rhinoceros (*Diceros bicornis*) from Cameroon spells not only the local extinction of rhinoceroses in Cameroon, but also the global extinction of the Western Black Rhinoceros (*Diceros bicornis longipes*).

In at least one case, scientists have found a local extinction useful for research: In the case of the Bay checkerspot butterfly, scientists, including Paul R. Ehrlich, chose not to intervene in a local extinction, using it to study the danger to the world population. However, similar studies are not carried out where a global population is at risk.

IUCN Subpopulation and Stock Assessments

While the World Conservation Union (IUCN) mostly only categorizes whole species or subspecies, assessing the global risk of extinction, in some cases it also assesses the risks to stocks and populations, especially to preserve genetic diversity. In all, 119 stocks or subpopulations across 69 species have been assessed by the IUCN in 2006.

Examples of stocks and populations assessed by the IUCN for the threat of local extinction:

- Marsh deer (three subpopulations assessed)

- Blue whale, North Pacific stock and North Atlantic stock

- Bowhead whale, *Balaena mysticetus* (five subpopulations assessed), from Critically Endangered to LR/cd

- Lake sturgeon, *Acipenser fulvescens*, Mississippi & Missouri Basins subpopulation assessed as Vulnerable

- Wild common carp, *Cyprinus carpio* (River Danube subpopulation)

- Black-flanked rock-wallaby *Petrogale lateralis* (MacDonnell Ranges subpopulation and Western Kimberly subpopulation)

The IUCN also lists countries where assessed species, subspecies or subpopulations are found, and from which countries they have been extirpated or reintroduced.

The IUCN has only three entries for subpopulations that have become extinct the Aral Sea stock of Ship sturgeon (*Acipenser nudiventris*); the Adriatic Sea stock of Beluga (*Huso huso*); and the Mexican subpopulation of Wolf (*Canis lupus*), which is extinct in the wild. No plant or fungi subpopulations have been assessed by the IUCN.

Local Extinction Events

Major environmental events, such as volcanic eruptions, may lead to large numbers of local extinctions, such as with the 1980 Mount St. Helens eruption, which led to a fern spike extinction.

Background Extinction Rate

Background extinction rate, also known as the normal extinction rate, refers to the standard rate of extinction in earth's geological and biological history before humans became a primary contributor to extinctions. This is primarily the pre-human extinction rates during periods in between major extinction events.

Overview

Extinctions are a normal part of the evolutionary process, and the background extinction rate is a measurement of "how often" they naturally occur. Normal extinction rates are often used as a comparison to present day extinction rates, to illustrate the higher frequency of extinction today than in all periods of non-extinction events before it.

Background extinction rates have not remained constant, although changes are measured over geological time, covering millions of years.

Measurement

Background extinction rates are typically measured three different ways. The first is simply the number of species that normally go extinct over a given period of time. For example, at the background rate one species of bird will go extinct every estimated 400 years. Another way the extinction rate can be given is in million species years (MSY). For example, there is approximately one extinction estimated per million species years. From a purely mathematical standpoint this means that if there are a million species on the planet earth, one would go extinct every year, while if there was only one species it would go extinct in one million years, etc. The third way is in giving species survival rates over time. For example, given normal extinction rates species typically exist for 5–10 hundred thousand years before going extinct.

Lifespan Estimates

Some species lifespan estimates by taxonomy

Taxonomy	Source of Estimate	Species Average Lifespan (Millions of Years)
All Invertebrates	Raup (1978)	11
Marine Invertebrates	Valentine (1970)	5–10
Marine Animals	Raup (1991)	4
Marine Animals	Sepkoski (1992)	5
All Fossil Groups	Simpson (1952)	0.5–5
Mammals	Martin (1993)	1
Cenozoic Mammals	Raup and Stanley (1978)	1–2
Diatoms	Van Valen	8
Dinoflagelates	Van Valen (1973)	13
Planktonic Foraminifera	Van Valen (1973)	7
Cenozoic Bivalves	Raup and Stanley (1978)	10
Echinoderms	Durham (1970)	6
Silurian Graptolites	Rickards (1977)	2

Accuracy

The fact that we do not currently know the total number of species, in the past nor the present, makes it very difficult to accurately calculate the non-anthropogenically influenced extinction rates. As a rate, it is essential to know not just the number of extinctions, but also the number of non-extinctions. This fact, coupled with the fact that the rates do not remain constant, significantly reduces accuracy in estimates of the normal rate of extinctions.

Extinction Threshold

Extinction threshold is a term used in conservation biology to explain the point at which a species, population or metapopulation, experiences an abrupt change in density or number because of an important parameter, such as habitat loss. It is at this critical value below which a species, population, or metapopulation, will go extinct, though this may take a long time for species just below the critical value, a phenomenon known as extinction debt.

Extinction thresholds are important to conservation biologists when studying a species in a population or metapopulation context because the colonization rate must be larger than the extinction rate, otherwise the entire entity will go extinct once it reaches the threshold.

Extinction thresholds are realized under a number of circumstances and the point in modeling them is to define the conditions that lead a population to extinction. Modeling extinction thresholds can explain the relationship between extinction threshold and habitat loss and habitat fragmentation.

Mathematical Models

Metapopulation-type models are used to predict extinction thresholds. The classic metapopulation model is the Levins Model, which is the model of metapopulation dynamics established by Richard Levins in the 1960s. It was used to evaluate patch occupancy in a large network of patches. This model was extended in the 1980s by Russell Lande to include habitat occupancy. This mathematical model is used to infer the extinction values and important population densities. These mathematical models are primarily used to study extinction thresholds because of the difficulty in understanding extinction processes through empirical methods and the current lack of research on this subject. When determining an extinction threshold there are two types of models that can be used: deterministic and stochastic metapopulation models.

Deterministic

Deterministic metapopulation models assume that there are an infinite number of habitat patches available and predict that the metapopulation will go extinct only if the threshold is not met.

$$dp/dt = chp\ (1\text{-}p)\text{-}ep$$

Where p= occupied patches, e= extinction rate, c= colonization rate, and h= amount of habitat.

A species will persist only if $h > \delta$

where $\delta = e/c$

δ= species parameter, or how successful a species is in colonizing a patch.

Stochastic

Stochastic metapopulation models take into account stochasticity, which is the non-deterministic or random processes in nature. With this approach a metapopulation may be above the threshold if determined that it is unlikely it will go extinct within a certain time period.

The complex nature of these models can result in a small metapopulation that is considered to be above the deterministic extinction threshold, but in reality has a high risk of extinction.

Other Factors

When using metapopulation-type models to predict extinction thresholds there are a number of factors that can affect the results of a model. First, including more complicated models, rather than relying solely on the Levins model produces different dynamics. For example, in an article published in 2004, Otso Ovaskainen and Ilkka Hanski explained with an empirical example that when factors such as Allee effect or Rescue effect were included in modeling the extinction threshold, there were unexpected extinctions in a high number of species. A more complex model came up with different results, and in practicing conservation biology this can add more confusion to efforts to save a species from the extinction threshold. Transient dynamics, which are effects on the extinction threshold because of instability in either the metapopulation or environmental conditions, is also a large player in modeling results. Landscapes that have recently endured habitat loss and fragmentation may be less able to sustain a metapopulation than previously understood without considering transient dynamics. Finally, environmental stochasticity, which may be spatially correlated, can lead to amplified regional stochastic fluctuations and therefore greatly affect the extinction risk.

Latent Extinction Risk

In conservation biology, latent extinction risk is a measure of the potential for a species to become threatened.

Latent risk can most easily be described as the difference, or discrepancy, between the current observed extinction risk of a species (typically as quantified by the [IUCN Red List]) and the theoretical extinction risk of a species predicted by its biological or life history characteristics.

Calculation

Because latent risk is the discrepancy between current and predicted risks, estimates of both of these values are required. Once these values are known, the latent extinction risk can be calculated as *Predicted Risk - Current Risk = Latent Extinction Risk.*

When the latent extinction risk is a positive value, it indicates that a species is currently less threatened than its biology would suggest it ought to be. For example, a species may have several of the characteristics often found in threatened species, such as large body size, small geographic distribution, or low reproductive rate, but still be rated as "least concern" in the IUCN Red List. This may be because it has not yet been exposed to serious threatening processes such as habitat degradation.

Conversely, negative values of latent risk indicate that a species is already more threatened than its biology would indicate, probably because it inhabits a part of the world where it has been exposed to extreme endangering processes. Species with severely low negative values are usually listed as an endangered species and have associated recovery and conservation plans.

Limits

One of the issues associated with latent extinction risk is its difficulty to calculate because of the limited availability of data for predicting extinction risk across large numbers of species. Hence, the only study of latent risk to date has focused on mammals, which are one of the best-studied groups of organisms.

Effects on Conservation

A study of latent extinction risk in mammals identified a number of "hotspots" where the average value of latent risk for mammal species was unusually high. This study suggested that these areas represented an opportunity for proactive conservation efforts, because these could become the "future battlegrounds of mammal conservation" if levels of human impact increase. Unexpectedly, the hotspots of mammal latent risk include large areas of Arctic America, where overall mammal diversity is not high, but where many species have the kind of biological traits (such as large body size and slow reproductive rate) that could render them extinction-prone. Another notable region of high latent risk for mammals is the island chain of Indonesia and Melanesia, where there are large numbers of restricted-range endemic species.

Because it is much more cost-effective to prevent species declines before they happen than to attempt to rescue species from the brink of extinction, latent risk hotspots could form part of a global scheme to prioritize areas for conservation effort, together with other kinds of priority areas such as biodiversity hotspots.

Extinction Debt

In ecology, extinction debt is the future extinction of species due to events in the past. Extinction debt occurs because of time delays between impacts on a species, such as destruction of habitat, and the species' ultimate disappearance. For instance, long-lived trees may survive for many years even after reproduction of new trees has become impossible, and thus they may be committed to extinction. Technically, extinction debt generally refers to the *number of species* in an area likely to become extinct, rather than the prospects of any one species, but colloquially it refers to any occurrence of delayed extinction.

Extinction debt may be local or global, but most examples are local as these are easier to observe and model. It is most likely to be found in long-lived species and species with very specific habitat requirements (specialists). Extinction debt has important implications for conservation, as it implies that species may become extinct due to past habitat destruction, even if continued impacts cease, and that current reserves may not be sufficient to maintain the species that occupy them. Interventions such as habitat restoration may reverse extinction debt.

Immigration credit is the corollary to extinction debt. It refers to the number of species likely to immigrate to an area after an event such as the restoration of an ecosystem.

Terminology

The term "extinction debt" was first used in 1994 in a paper by David Tilman, Robert May, Clarence Lehman and Martin Nowak, although Jared Diamond used the term "relaxation time" to describe a similar phenomenon in 1972.

Extinction debt is also known by the terms dead clade walking, and survival without recovery when referring to the species affected. The phrase "dead clade walking" was coined by David Jablonski as early as 2001 as a reference to *Dead Man Walking*, a film whose title is based on American prison slang for a condemned prisoner's last walk to the execution chamber. "Dead clade walking" has since appeared in other scientists' writings about the aftermaths of mass extinctions.

In discussions of threats to biodiversity, extinction debt is analogous to the "climate commitment" in climate change, which states that inertia will cause the earth to continue to warm for centuries even if no more greenhouse gasses are emitted. Similarly, the current extinction may continue long after human impacts on species halt.

Causes

Jablonski recognized at least four patterns in the fossil record following mass extinctions: (1) unbroken continuity (large-scale patterns continuing with little disruption), (2) continuity with setbacks (patterns disturbed by the extinction event but soon continuing on the previous trajectory), (3) survival without recovery or "dead clade walking" (a group's dwindling to extinction or minor ecological niches), and (4) unbridled diversification (an increase in diversity and species richness, as in the mammals following the end-Cretaceous extinction event).

Extinction debt is caused by many of the same drivers as extinction. The most well-known drivers of extinction debt are habitat fragmentation and habitat destruction. These cause extinction debt by reducing the ability of species to persist via immigration to new habitats. Under equilibrium conditions, species may become extinct in one habitat patch, yet continues to survive because it can disperse to other patches. However, as other patches have been destroyed or rendered inaccessible due to fragmentation, this "insurance" effect is reduced and the species may ultimately become extinct.

Pollution may also cause extinction debt by reducing a species' birth rate or increasing its death rate so that its population slowly declines. Extinction debts may be caused by invasive species or by climate change.

Extinction debt may also occur due to the loss of mutualist species. In New Zealand, the local extinction of several species of pollinating birds in 1870 has caused a long-term reduction in the reproduction of the shrub species *Rhabdothamnus solandri*, which requires these birds to produce seeds. However, as the plant is slow-growing and long-lived, its populations persist.

Jablonski found that the extinction rate of marine invertebrates was significantly higher in the stage (major subdivision of an epoch – typically 2–10 million years' duration) following a mass extinction than in the stages preceding the mass extinction. His analysis focused on marine molluscs since they constitute the most abundant group of fossils and are therefore the least likely to produce sampling errors. Jablonski suggested that two possible explanations deserved further study:

- Post-extinction physical environments differed from pre-extinction environments in ways which were disadvantageous to the "dead clades walking".

- Ecosystems that developed after recoveries from mass extinctions may have been less favorable for the "dead clades walking".

Time Scale

The time to "payoff" of extinction debt can be very long. Islands that lost habitat at the end of the last ice age 10,000 years ago still appear to be losing species as a result. It has been shown that some bryozoans, a type of microscopic marine organism, became extinct due to the volcanic rise of the Isthmus of Panama. This event cut off the flow of nutrients from the Pacific Ocean to the Caribbean 3–4.5 million years ago. While bryozoan populations dropped severely at this time, extinction of these species took another 1–2 million years.

Extinction debts incurred due to human actions have shorter timescales. Local extinction of birds from rainforest fragmentation occurs over years or decades, while plants in fragmented grasslands show debts lasting 50 to 100 years. Tree species in fragmented temperate forests have debts lasting 200 years or more.

Theoretical Development

Origins in Metapopulation Models

Tilman et al. demonstrated that extinction debt could occur using a mathematical ecosystem model of species metapopulations. Metapopulations are multiple populations of a species that live in separate habitat patches or islands but interact via immigration between the patches. In this model, species persist via a balance between random local extinctions in patches and colonization of new patches. Tilman et al. al used this model to predict that species would persist long after they no longer had sufficient habitat to support them. When used to estimate extinction debts of tropical tree species, the model predicted debts lasting 50–400 years.

One of the assumptions underlying the original extinction debt model was a trade-off between species' competitive ability and colonization ability. That is, a species that competes well against other species, and is more likely to become dominant in an area, is less likely to colonize new habitats due to evolutionary trade-offs . One of the implications of this assumption is that better competitors, which may even be more common than other species, are more likely to become extinct than rarer, less competitive, better dispersing species. This has been one of the more controversial components of the model, as there is little evidence for this trade-off in many ecosystems, and in many empirical studies dominant competitors were least likely species to become extinct. A later modification of the model showed that these trade-off assumptions may be relaxed, but need to exist partially, in order for the theory to work.

Development in Other Models

Further theoretical work has shown that extinction debt can occur under many different circumstances, driven by different mechanisms and under different model assumptions. The original model predicted extinction debt as a result of habitat destruction in a system of small, isolated habitats such as islands. Later models showed that extinction debt could occur in systems where habitat destruction occurs in small areas within a large area of habitat, as in slash-and-burn agriculture in forests, and could also occur due to decreased growth of species from pollutants. Predicted patterns of extinction debt differ between models, though. For instance, habitat destruction resembling slash-and-burn agriculture is thought to affect rare species rather than poor colonizers. Models that incorporate stochasticity, or random fluctuation in populations, show extinction debt occurring over different time scales than classic models.

Most recently, extinction debts have been estimated through the use models derived from neutral theory. Neutral theory has very different assumptions than the metapopulation models described above. It predicts that the abundance and distribution of species can be predicted entirely through random processes, without considering the traits of individual species. As extinction debt arises in models under such different assumptions, it is robust to different kinds of models. Models derived from neutral theory have successfully predicted extinction times for a number of bird species, but perform poorly at both very small and very large spatial scales.

Mathematical models have also shown that extinction debt will last longer if it occurs in response to large habitat impacts (as the system will move farther from equilibrium), and if species are long-lived. Also, species just below their extinction threshold, that is, just below the population level or habitat occupancy levels required sustain their population, will have long-term extinction debts. Finally, extinction debts are predicted to last longer in landscapes with a few large patches of habitat, rather than many small ones.

Detection

Extinction debt is difficult to detect and measure. Processes that drive extinction debt are inherently slow and highly variable (noisy), and it is difficult to locate or count the very small populations of near-extinct species. Because of these issues, most measures of extinction debt have a great deal of uncertainty.

Experimental Evidence

Due to the logistical and ethical difficulties of inciting extinction debt, there are few studies of extinction debt in controlled experiments. However, experiments microcosms of insects living on moss habitats demonstrated that extinction debt occurs after habitat destruction. In these experiments, it took 6–12 months for species to die out following the destruction of habitat.

Observational Methods

Long-term Observation

Extinction debts that reach equilibrium in relatively short time scales (years to decades) can be observed via measuring the change in species numbers in the time following an impact on habitat.

For instance, in the Amazon rainforest, researchers have measured the rate at which bird species disappear after forest is cut down. As even short-term extinction debts can take years to decades to reach equilibrium, though, such studies take many years and good data are rare.

Comparing The Past and Present

Most studies of extinction debt compare species numbers with habitat patterns from the past and habitat patterns in the present. If the present populations of species are more closely related to past habitat patterns than present, extinction debt is a likely explanation. The magnitude of extinction debt (i.e., number of species likely to become extinct) can not be estimated by this method.

If one has information on species populations from the past in addition to the present, the magnitude of extinction debt can be estimated. One can use the relationship between species and habitat from the past to predict the number of species expected in the present. The difference between this estimate and the actual number of species is the extinction debt.

This method requires the assumption that in the past species and their habitat were in equilibrium, which is often unknown. Also, a common relationship used to equate habitat and species number is the species-area curve, but as the species-area curve arises from very different mechanisms than those in metapopulation based models, extinction debts measured in this way may not conform with metapopulation models' predictions. The relationship between habitat and species number can also be represented by much more complex models that simulate the behavior of many species independently.

Comparing Impacted and Pristine Habitats

If data on past species numbers or habitat are not available, species debt can also be estimated by comparing two different habitats: one which is mostly intact, and another which has had areas cleared and is smaller and more fragmented. One can then measure the relationship of species with the condition of habitat in the intact habitat, and, assuming this represents equilibrium, use it to predict the number of species in the cleared habitat. If this prediction is lower than the actual number of species in the cleared habitat, then the difference represents extinction debt. This method requires many of the same assumptions as methods comparing the past and present.

Examples

Grasslands

Studies of European grasslands show evidence of extinction debt through both comparisons with the past and between present-day systems with different levels of human impacts. The species diversity of grasslands in Sweden appears to be a remnant of more connected landscapes present 50 to 100 years ago. In alvar grasslands in Estonia that have lost area since the 1930s, 17-70% of species are estimated to be committed to extinction. However, studies of similar grasslands in Belgium, where similar impacts have occurred, show no evidence of extinction debt. This may be due to differences in the scale of measurement or the level of specialization of grass species.

Forests

Forests in Vlaams-Brabant, Belgium, show evidence of extinction debt remaining from deforestation that occurred between 1775 and 1900. Detailed modeling of species behavior, based on similar forests in England that did not experience deforestation, showed that long-lived and slow-growing species were more common than equilibrium models would predict, indicating that their presence was due to lingering extinction debt.

In Sweden, some species of lichens show an extinction debt in fragments of ancient forest. However, species of lichens that are habitat generalists, rather than specialists, do not.

Insects

Extinction debt has been found among species of butterflies living in the grasslands on Saaremaa and Muhu - islands off the western coast of Estonia. Butterfly species distributions on these islands are better explained by the habitat in the past than current habitats.

On the islands of the Azores Archipelago, more than 95% of native forests have been destroyed in the past 600 years. As a result, more than half of arthropods on these islands are believed to be committed to extinction, with many islands likely to lose more than 90% of species.

Vertebrates

80-90% of extinction from past deforestation in the Amazon has yet to occur, based on modeling based on species-area relationships. Local extinctions of approximately 6 species are expected in each 2500 km² region by 2050 due to past deforestation. Birds in the Amazon rain forest continued to become extinct locally for 12 years following logging that broke up contiguous forest into smaller fragments. The extinction rate slowed, however, as forest regrew in the spaces in between habitat fragments.

Countries in Africa are estimated to have, on average, a local extinction debt of 30% for forest-dwelling primates. That is, they are expected to have 30% of their forest primate species to become extinct in the future due to loss of forest habitat. The time scale for these extinctions has not been estimated.

Based on historical species-area relationships, Hungary currently has approximately nine more species of raptors than are thought to be able to be supported by current nature reserves.

Applications to Conservation

The existence of extinction debt in many different ecosystems has important implications for conservation. It implies that in the absence of further habitat destruction or other environmental impacts, many species are still likely to become extinct. Protection of existing habitats may not be sufficient to protect species from extinction. However, the long time scales of extinction debt may allow for habitat restoration in order to prevent extinction, as occurred in the slowing of extinction in Amazon forest birds above. In another example, it has been found that grizzly bears in very small reserves in the Rocky Mountains are likely to become extinct, but this finding allows the modification of reserve networks to better support their populations.

The extinction debt concept may require revision of the value of land for species conservation, as the number of species currently present in a habitat may not be a good measure of the habitat's ability to support species in the future. As extinction debt may last longest near extinction thresholds, it may be hardest to detect the threat of extinction for species that conservation could benefit the most.

Economic analyses have shown that including extinction in management decision-making process changes decision outcomes, as the decision to destroy habitat changes conservation value in the future as well as the present. It is estimated that in Costa Rica, ongoing extinction debt may cost between $88 million and $467 million.

Extinction Risk from Global Warming

The extinction risk of global warming is the risk of species becoming extinct due to the effects of global warming.

Current Projections

The scientific consensus in the IPCC Fourth Assessment Report is that

"Anthropogenic warming could lead to some impacts that are abrupt or irreversible, depending upon the rate and magnitude of the climate change."

"There is medium confidence that approximately 20-30% of species assessed so far are likely to be at increased risk of extinction if increases in global average warming exceed 1.5-2.5 °C (relative to 1980-1999). As global average temperature increase exceeds about 3.5 °C, model projections suggest significant extinctions (40-70% of species assessed) around the globe."

In one study published in *Nature* in 2004, between 15 and 37% of 1103 endemic or near-endemic known plant and animal species will be "committed to extinction" by 2050. More properly, changes in habitat by 2050 will put them outside the survival range for the inhabitants, thus committing the species to extinction.

Other researchers, such as Thuiller *et al.*, Araújo *et al.*, Person *et al.*, Buckley and Roughgarden, and Harte *et al.* have raised concern regarding uncertainty in Thomas *et al.*'s projections; some of these studies believe it is an overestimate, others believe the risk could be greater. Thomas *et al.* replied in Nature addressing criticisms and concluding "Although further investigation is needed into each of these areas, it is unlikely to result in substantially reduced estimates of extinction. Anthropogenic climate change seems set to generate very large numbers of species-level extinctions." On the other hand, Daniel Botkin *et al.* state "... global estimates of extinctions due to climate change (Thomas et al. 2004) may have greatly overestimated the probability of extinction..."

Mechanistic studies are documenting extinctions due to recent climate change: McLaughlin *et al.* documented two populations of Bay checkerspot butterfly being threatened by precipitation change. Parmesan states, "Few studies have been conducted at a scale that encompasses an entire species" and McLaughlin *et al.* agreed "few mechanistic studies have linked extinctions to recent climate change."

In 2008, the white lemuroid possum was reported to be the first known mammal species to be driven extinct by man-made global warming. However, these reports were based on a misunderstanding. One population of these possums in the mountain forests of northern Queensland is severely threatened by climate change as the animals cannot survive extended temperatures over 30 °C. However, another population 100 kilometres south remains in good health.

According to research published in the January 4, 2012 *Proceedings of the Royal Society B* current climate models may be flawed because they overlook two important factors: the differences in how quickly species relocate and competition among species. According to the researchers, led by Mark C. Urban, an ecologist at the University of Connecticut, diversity decreased when they took these factors into account, and that new communities of organisms, which do not exist today, emerged. As a result the rate of extinctions may be higher than previously projected.

According to research published in the 30 May 2014 issue of *Science,* most known species have small ranges, and the numbers of small-ranged species are increasing quickly. They are geographically concentrated and are disproportionately likely to be threatened or already extinct. According to the research, current rates of extinction are three orders of magnitude higher than the background extinction rate, and future rates, which depend on many factors, are poised to increase. Although there has been rapid progress in developing protected areas, such efforts are not ecologically representative, nor do they optimally protect biodiversity. In the researchers' view, human activity tends to destroy critical habitats where species live, warms the planet, and tends to move species around the planet to places where they don't belong and where they can come into conflict with human needs (e.g. causing species to become pests).

In 2016 the Bramble Cay melomys, which lived on a Great Barrier Reef island, was reported to probably be the first mammal to become extinct because of sea level rises due to human-made climate change.

References

- Newman, Mark (1997). "A model of mass extinction". Journal of Theoretical Biology. 189: 235–252. doi:10.1006/jtbi.1997.0508

- Mills, L. Scott (2009-03-12). Conservation of Wildlife Populations: Demography, Genetics and Management. John Wiley & Sons. p. 13. ISBN 9781444308938

- Quince, C.; et al. "Deleting species from model food webs" (PDF). Archived from the original (PDF) on 2006-09-25. Retrieved 2007-02-15

- Buckley, L. B; Roughgarden (2004). "Biodiversity conservation: Effects of changes in climate and land use". Nature. 430 (6995). doi:10.1038/nature02717

- Leroux, A. D.; Martin, V. L.; Goeschl, T. (2009). "Optimal conservation, extinction debt, and the augmented quasi-option value". Journal of Environmental Economics and Management. 58: 43. doi:10.1016/j.jeem.2008.10.002

- Stearns, Beverly Peterson and Stephen C. (2000). "Preface". Watching, from the Edge of Extinction. Yale University Press. pp. x. ISBN 0-300-08469-2

- Graham, Chris (July 11, 2017). "Earth undergoing sixth 'mass extinction' as humans spur 'biological annihilation' of wildlife". The Telegraph. Retrieved July 23, 2017

- Thuiller, W.; Araújo, M.B.; Pearson, R.G.; Whittaker, R.J.; Brotons, L.; Lavorel, S. (2004). "Biodiversity conservation: Uncertainty in predictions of extinction risk". Nature. 430 (6995): 1. doi:10.1038/nature02716

- Etienne, R.; Nagelkerke, C. (2002). "Non-equilibria in Small Metapopulations: Comparing the Deterministic Levins Model with its Stochastic Counterpart". Journal of Theoretical Biology. 219 (4): 463–78. PMID 12425979. doi:10.1006/jtbi.2002.3135

- Clover, Charles (2004). The End of the Line: How overfishing is changing the world and what we eat. London: Ebury Press. ISBN 0-09-189780-7

- Baldi, A.; Voros, J. (2006). "Extinction debt of Hungarian reserves: A historical perspective". Basic and Applied Ecology. 7: 289. doi:10.1016/j.baae.2005.09.005

- McKenzie, N. L.; Burbidge, A. A.; Baynes, A.; Brereton, R. N.; Dickman, C. R.; Gordon, G.; Gibson, L. A.; Menkhorst, P. W.; et al. (2007), "Analysis of factors implicated in the recent decline of Australia's mammal fauna", Journal of Biogeography, 34 (4): 597–611, doi:10.1111/j.1365-2699.2006.01639.x

- Bradshaw, Corey JA; Sodhi, Navjot S; Brook, Barry W (2009). "Tropical turmoil: a biodiversity tragedy in progress". Frontiers in Ecology and the Environment. 7: 79–87. doi:10.1890/070193

- Ehrlich, Anne (1981). Extinction: The Causes and Consequences of the Disappearance of Species. Random House, New York. ISBN 0-394-51312-6

- Botkin, Daniel B.; et al. (March 2007). "Forecasting the Effects of Global Warming on Biodiversity" (PDF). BioScience. 57 (3): 227–236. doi:10.1641/B570306. Retrieved 2007-11-30

- Lindborg, R.; Eriksson, O. (2004). "Historical Landscape Connectivity Affects Present Plant Species Diversity". Ecology. 85: 1840. doi:10.1890/04-0367

4

An Integrated Study of Wildlife and its Conservation

Animals which are not domesticated are a part of wildlife. A broader definition would be organisms that are found in the wild. It includes various types of animals, plants and fungi. Wildlife trade, wildlife conservation, conservation biology, habitat conservation and conservation movement are some of the topics related to wildlife conservation. The aspects elucidated in this chapter are of vital importance, and provide a better understanding of wildlife habitat management.

Wildlife

Tiger *Panthera tigris*

Wildlife traditionally refers to undomesticated animal species, but has come to include all plants, fungi, and other organisms that grow or live wild in an area without being introduced by humans.

Wildlife can be found in all ecosystems. Deserts, forests, rain forests, plains, grasslands and other areas including the most developed urban areas, all have distinct forms of wildlife. While the term in popular culture usually refers to animals that are untouched by human factors, most scientists agree that much wildlife is affected by human activities.

Humans have historically tended to separate civilization from wildlife in a number of ways including the legal, social, and moral sense. Some animals, however, have adapted to suburban environments. This includes such animals as domesticated cats, dogs, mice, and gerbils. Some religions declare certain animals to be sacred, and in modern times concern for the natural environment has provoked activists to protest against the exploitation of wildlife for human benefit or entertainment.

The global wildlife population has decreased by 52 percent between 1970 and 2014, according to a report by the World Wildlife Fund.

Food, Pets, and Traditional Medicines

A mesh bag full of live frogs waiting for a buyer at Chiang Mai's Thanin market.
Frog meat in Thailand is mostly used in stir-fries and Thai curries.

Anthropologists believe that the Stone Age people and hunter-gatherers relied on wildlife, both plants and animals, for their food. In fact, some species may have been hunted to extinction by early human hunters. Today, hunting, fishing, and gathering wildlife is still a significant food source in some parts of the world. In other areas, hunting and non-commercial fishing are mainly seen as a sport or recreation, with the edible meat as mostly a side benefit of it. Meat sourced from wildlife that is not traditionally regarded as game is known as bush meat. The increasing demand for wildlife as a source of traditional food in East Asia is decimating populations of sharks, primates, pangolins and other animals, which they believe have aphrodisiac properties.

In November 2008, almost 900 plucked and "oven-ready" owls and other protected wildlife species were confiscated by the Department of Wildlife and National Parks in Malaysia, according to TRAFFIC. The animals were believed to be bound for China, to be sold in wild meat restaurants. Most are listed in CITES (the Convention on International Trade in Endangered Species of Wild Fauna and Flora) which prohibits or restricts such trade.

> 66   Malaysia is home to a vast array of amazing wildlife. However, illegal hunting and   99
>   trade poses a threat to Malaysia's natural diversity.

A November 2008 report from biologist and author Sally Kneidel, PhD, documented numerous wildlife species for sale in informal markets along the Amazon River, including wild-caught marmosets sold for as little as $1.60 (5 Peruvian soles). Many Amazon species, including peccaries, agoutis, turtles, turtle eggs, anacondas, armadillos, etc., are sold primarily as food. Others in these informal markets, such as monkeys and parrots, are destined for the pet trade, often smuggled into the United States. Still other Amazon species are popular ingredients in traditional medicines sold in local markets. The medicinal value of animal parts is based largely on superstition.

Religion

Many animal species have spiritual significance in different cultures around the world, and they and their products may be used as sacred objects in religious rituals. For example, eagles, hawks and their feathers have great cultural and spiritual value to Native Americans as religious objects. In Hinduism the cow is regarded sacred.

Muslims conduct sacrifices on Eid-ul-Adha to commemorate the sacrificial spirit of Ibrahim [Abraham] in love of God. Camels, sheep, goats, and cows may be offered as sacrifice during the three days of Eid.

Tourism

Many nations have established their tourism sector around their natural wildlife. South Africa has, for example, many opportunities for tourists to see the country's wildlife in its national parks, such as the Kruger Park. In South India the Periar Wildlife Sanctuary, Bandipur National Park and Mudamalai Wildlife Sanctuary are situated around and in forests. India is home to many national parks and wildlife sanctuaries showing the diversity of its wildlife, much of its unique fauna, and excels in the range. There are 89 national parks, 13 bio reserves and more than 400 wildlife sanctuaries across India which are the best places to go to see tigers, lions, elephants, rhinoceros, birds, and other wildlife which reflect the importance that the country places on nature and wildlife conservation.

Destruction

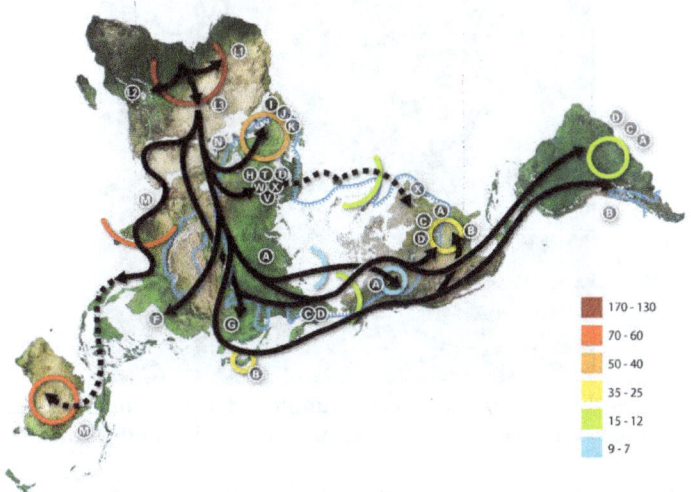

Map of early human migrations, according to mitochondrial population genetics.
Numbers are millennia before the present.

This section focuses on anthropogenic forms of wildlife destruction.

Exploitation of wild populations has been a characteristic of modern man since our exodus from Africa 130,000 – 70,000 years ago. The rate of extinctions of entire species of plants and animals across the planet has been so high in the last few hundred years it is widely believed that we are in the sixth great extinction event on this planet; the Holocene Mass Extinction.

Destruction of wildlife does not always lead to an extinction of the species in question, however, the dramatic loss of entire species across Earth dominates any review of wildlife destruction as extinction is the level of damage to a wild population from which there is no return.

The four most general reasons that lead to destruction of wildlife include overkill, habitat destruction and fragmentation, impact of introduced species and chains of extinction.

Overkill

Wildlife is an invaluable treasure but it is being exploited due to illegal trade of many of its species.Overkill happens whenever hunting occurs at rates greater than the reproductive capacity of the population is being exploited. The effects of this are often noticed much more dramatically in slow growing populations such as many larger species of fish. Initially when a portion of a wild population is hunted, an increased availability of resources (food, etc.) is experienced increasing growth and reproduction as density dependent inhibition is lowered. Hunting, fishing and so on, has lowered the competition between members of a population. However, if this hunting continues at rate greater than the rate at which new members of the population can reach breeding age and produce more young, the population will begin to decrease in numbers.

Populations that are confined to islands, whether literal islands or just areas of habitat that are effectively an "island" for the species concerned, have also been observed to be at greater risk of dramatic population declines following unsustainable hunting.

Habitat Destruction and Fragmentation

Deforestation and increased road-building in the Amazon Rainforest are a significant concern because of increased human encroachment upon wild areas, increased resource extraction and further threats to biodiversity.

The habitat of any given species is considered its preferred area or territory. Many processes associated with human habitation of an area cause loss of this area and decrease the carrying capacity of the land for that species. In many cases these changes in land use cause a patchy break-up of the wild landscape. Agricultural land frequently displays this type of extremely fragmented, or relictual, habitat. Farms sprawl across the landscape with patches of uncleared woodland or forest dotted in-between occasional paddocks.

Examples of habitat destruction include grazing of bushland by farmed animals, changes to natural fire regimes, forest clearing for timber production and wetland draining for city expansion.

Impact of Introduced Species

Mice, cats, rabbits, dandelions and poison ivy are all examples of species that have become invasive threats to wild species in various parts of the world. Frequently species that are uncommon in their home range become out-of-control invasions in distant but similar climates. The reasons for this have not always been clear and Charles Darwin felt it was unlikely that exotic species would ever be able to grow abundantly in a place in which they had not evolved. The reality is that the vast majority of species exposed to a new habitat do not reproduce successfully. Occasionally, however, some populations do take hold and after a period of acclimation can increase in numbers significantly, having destructive effects on many elements of the native environment of which they have become part.

Chains of Extinction

This final group is one of secondary effects. All wild populations of living things have many complex intertwining links with other living things around them. Large herbivorous animals such as the hippopotamus have populations of insectivorous birds that feed off the many parasitic insects that grow on the hippo. Should the hippo die out, so too will these groups of birds, leading to further destruction as other species dependent on the birds are affected. Also referred to as a domino effect, this series of chain reactions is by far the most destructive process that can occur in any ecological community.

Another example is the black drongos and the cattle egrets found in India. These birds feed on insects on the back of cattle, which helps to keep them disease-free. Destroying the nesting habitats of these birds would cause a decrease in the cattle population because of the spread of insect-borne diseases.

Media

The Douglas Squirrel

Wildlife has long been a common subject for educational television shows. National Geographic specials appeared on CBS beginning in 1965, later moving to ABC and then PBS. In 1963, NBC debuted *Wild Kingdom,* a popular program featuring zoologist Marlin Perkins as host. The BBC natural history unit in the UK was a similar pioneer, the first wildlife series LOOK presented by Sir Peter Scott, was a studio-based show, with filmed inserts. It was in this series that David Attenborough first made his appearance which led to the series Zoo Quest during which he and cameraman Charles Lagus went to many exotic places looking for and filming elusive wildlife—notably the Komodo dragon in Indonesia and lemurs in Madagascar. Since 1984, the Discovery Channel and its spin off Animal Planet in the US have dominated the market for shows about wildlife on cable television, while on PBS the NATURE strand made by WNET-13 in New York and NOVA by WGBH in Boston are notable. Wildlife television is now a multimillion-dollar industry with specialist documentary film-makers in many countries including UK, US, New Zealand NHNZ, Australia, Austria, Germany, Japan, and Canada. There are many magazines which cover wildlife including National Wildlife Magazine, Birds & Blooms, Birding (magazine), and Ranger Rick (for children).

Wildlife Corridor

A wildlife corridor, habitat corridor, or green corridor is an area of habitat connecting wildlife populations separated by human activities or structures (such as roads, development, or logging). This allows an exchange of individuals between populations, which may help prevent the negative effects of inbreeding and reduced genetic diversity (via genetic drift) that often occur within isolated populations. Corridors may also help facilitate the re-establishment of populations that have been reduced or eliminated due to random events (such as fires or disease).

This may potentially moderate some of the worst effects of habitat fragmentation, wherein urbanization can split up habitat areas, causing animals to lose both their natural habitat and the ability to move between regions to use all of the resources they need to survive. Habitat fragmentation due to human development is an ever-increasing threat to biodiversity, and habitat corridors are a possible mitigation.

Purpose

The main goal of implementing habitat corridors is to increase biodiversity. When areas of land are broken up by human interference, population numbers become unstable and many animal and plant species become endangered. By re-connecting the fragments, the population fluctuations can decrease dramatically. Corridors can contribute to three factors that stabilize a population:

- Colonization—animals are able to move and occupy new areas when food sources or other natural resources are lacking in their core habitat.

- Migration—species that relocate seasonally can do so more safely and effectively when it does not interfere with human development barriers.

- Interbreeding—animals can find new mates in neighboring regions so that genetic diversity can increase and thus have a positive impact on the overall population.

Although corridors have been implemented with the assumption that they will increase biodiversity, not enough research has been done to come to a solid conclusion. The case for corridors has been built more on intuition and much less on empirical evidence (Tewksbury et al. 2002). Another factor that needs to be taken into account is what species the corridor is intended for. Some species have reacted more positively to corridors than others.

A habitat corridor could be considered as a possible solution in an area where destruction of a natural area has greatly affected its native species. Development such as roads, buildings, and farms can interrupt plants and animals in the region being destroyed. Furthermore, natural disasters such as wildfires and floods can leave animals with no choice but to evacuate. If the habitat is not connected to a safer one, it will ultimately lead to death. A remaining portion of natural habitat is called a remnant, and such portions need to be connected, because when migration decreases, extinction increases (Fleury 1997).

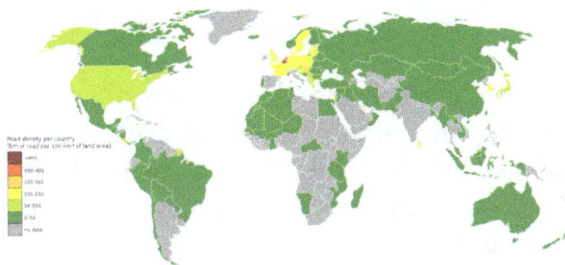

Map indicating the amount of roads per 100 km² of land area for each country

Corridors can be made in two distinct areas—either water or land. Water corridors are called riparian ribbons and usually come in the form of rivers and streams. Land corridors come on a scale as large as wooded strips connecting larger woodland areas. However, they can also be as simple as a line of shrubs along a sidewalk (Fleury 1997). Such areas can facilitate the movement of small animals, especially birds, from tree to tree, until they find a safe habitat to nest in. Not only do minimal corridors aid in the movement of animals, they are also aesthetically pleasing, which can sometimes encourage the community to accept and support them.

Users

Species can be categorized in one of two groups; passage users and corridor dwellers.

Passage users occupy corridors for brief periods of time. These animals use corridors for such events as seasonal migration, dispersal of a juvenile, or moving between parts of a large home range. Usually large herbivores, medium to large carnivores, and migratory species are passage users (Beier & Loe 1992). One common misconception is that the corridor only needs to be wide enough for the passage users to get through. However, the corridor still must be wide enough to be safe and also encourage the animals to use it, even though they do not live out their entire lives in it.

Corridor dwellers can occupy the passage anywhere from several days to several years. Species such as plants, reptiles, amphibians, birds, insects, and small mammals can spend their entire lives in linear habitats. In this case, the corridor must include everything that a species needs to live and breed, such as soil for germination, burrowing areas, and multiple other breeding adults (Beier & Loe 1992).

Types

Habitat corridors can be categorized according to their width. Typically the wider the corridor, the more use it will get from species. However, the width-length ratio, as well as design and quality play just as important of a role in creating the perfect corridor (Fleury 1997). The strip of land will suffer less from edge effects such as weeds, predators, and chemicals if it is constructed properly. The following are three divisions in corridor widths:

- Regional – (>500m wide); connect major ecological gradients such as migratory pathways.

- Sub-regional – (>300m wide); connect larger vegetated landscape features such as ridge-lines and valley floors.

- Local – (some <50m); connect remnant patches of gullies, wetlands, ridgelines, etc.

Habitat corridors can also be divided according to their continuity. Continuous corridors are strips that are not broken up, while "stepping stone" corridors are small patches of suitable habitat. When stepping stones are arranged in a line, they form a strip of land connecting two areas, just like a continuous corridor would. Both kinds provide linkages between protected core areas and stimulate or allow species to migrate.

Finally, corridors can come in the form of underpasses or overpasses, which can be very safe for both animals and humans. Many busy highways cross through natural habitats that native species occupy, as well. Large animals such as deer become a hazard when they cross in front of traffic and get hit. An overpass or an underpass serves as a bridge to facilitate the movement of animals across a busy road. Observations have shown that underpasses are actually more successful than overpasses because many times animals are too timid to cross over a bridge in front of traffic and would prefer to be more hidden (Dole et al. 2003).

Overpasses such as this one allow for traffic to continue for human convenience, while allowing wildlife to pass unharmed beneath from place to place.

Costs

Corridors can be expensive to plan out and put into action. For example, Daniel Simberloff et al. states that "a bridge that would maintain a riparian corridor costs about 13 times as much per lane-mile as would a road that would sever the corridor." He also states that maintenance of a corridor would be much more costly than refuges for endangered species. It would simply be easier to move animals between refuges than to buy land, install a corridor, and maintain it. However, where the goal is not just to preserve a few large animal species but to protect biodiversity among all plants and animals, then habitat corridors may be the only option. Corridors are going to be expensive to implement no matter what, but it does depend on the type, location, and size, which can all vary to a great degree. With the lack of field data on the effectiveness, many agencies are not willing to consider putting in corridors.

Monitoring Use

It is extremely important for researchers to pay attention to the population changes in animals after a corridor has been implemented to ensure that there are no harmful effects. Researchers can use both mark-recapture techniques and evaluate genetic flow in order to observe how much a corridor is being used. Marking and recapturing animals is more useful when keeping a close eye on individual movement (Mech & Hallet 2001). The only problem is that tagging animals and watching them does not tell anyone whether the migrating individuals are successfully mating with other populations in connected areas of land. On the other hand, genetic techniques can be more effective in evaluating migration and mating patterns.

One of the most important goals of developing a corridor is to increase migration in certain animal species. By looking at a population's gene flow, researchers can understand the genetic consequences of corridors (Mech & Hallett 2001). The migration patterns of an entire population are much more important than the movements of a few individuals. From these techniques, researchers will better understand whether or not habitat corridors are increasing biodiversity.

Stephen Mech and James Hallett introduce an additional reason genetic techniques are more useful; they "measure average migration rates over time, which reveals the effects of fragmentation of several generations and is not as sensitive to current population sizes as mark-recapture studies are." For example, when a population is extremely small, mark-recapture is almost impossible. Clearly, genetic analysis of a species is the best way to determine if animals are actually using corridors to move and reproduce.

Design

According to new research, wildlife corridors are best built with a certain degree of randomness or asymmetry, rather than built symmetrically. The research was conducted at UC Davis.

Wildlife corridors are susceptible to edge effects; habitat quality along the edge of a habitat fragment is often much lower than in core habitat areas. Wildlife corridors are important for large species requiring significant sized ranges; however, they are also vital as connection corridors for smaller animals and plants as well as ecological connectors to provide a *rescue effect*.

Examples

Both the safety of animals and humans can be achieved through the creation of corridors. For example, deer commonly cross roads in order to get to other grazing land. When they are faced with a car coming at them, they freeze; this puts both the deer and the human's life in danger. In Alberta, Canada, an overpass was constructed to keep animals off of the busy highway; the area is part of a national park, so many different creatures roam the area. The top of the bridge is covered in the native grass of the area so that it blends in better and animals will not know the difference. Gates were also put of on either side of the overpass to help guide animals in the right direction (Semrad 2007).

In Southern California, 15 underpasses and drainage culverts were observed to see how many animals used them as corridors. They proved to be especially effective on wide-ranging species such as carnivores, mule deer, small mammals, and reptiles, even though the corridors were not intended specifically for animals. Researchers also learned that factors such as surrounding habitat, underpass dimensions, and human activity also played a role in how much use they got. From this experiment, much was learned about what would constitute a successful habitat corridor (Dole et al. 2003).

In South Carolina, five remnant areas of land were monitored; one was put in the center and four were surrounding it. Then, a corridor was put between one of the remnants and the center. Butterflies that were placed in the center habitat were two to four times more likely to move to the connected remnant rather than the disconnected ones. Furthermore, male holly plants were placed in the center region, and female holly plants in the connected region increased by 70 percent in seed production compared to those plants in the disconnected region. The most impressive dispersal into the connected region, though, was through bird droppings. Far more plant seeds were dispersed through bird droppings in the corridor-connected patch of land (M. 2002).

There have also been positive effects on the rates of transfer and interbreeding in vole populations. A control population in which voles were confined to their core habitat with no corridor was compared to a treatment population in their core habitat with passages that they could use to move to other regions. Females typically stayed and mated within their founder population, but the rate of transfer through corridors in the males was very high. Researchers are not sure why the females did not move about as much, but it is apparent that the corridor effectively transferred at least some of the species to another location for breeding (Aars 1999).

In 2001, a wolf corridor was restored through a golf course in Jasper National Park, Alberta, which enabled wolves to pass through the course. After this restoration, wolves passed through the corridor frequently. This is one of the first demonstrations that corridors are used by wildlife, and can be effective in decreasing fragmentation. Earlier studies had been criticised for failing to demonstrate that corridor restoration leads to a change in wildlife behaviour.

Elephant corridors are narrow strips of land that allow elephants to move from one habitat patch to another. There are 88 identified elephant corridors in India.

In Africa, Botswana houses the largest number of free-roaming elephant herds. Elephants Without Borders (EWB) studies the movement of elephants is working to gain community support of local community corridors, so that elephants and humans can co-exist.

Major Wildlife Corridors

Several artificial wildlife corridors have been created, these include:

- the Paséo Pantera (also known as the MesoAmerican Biological corridor or Paséo del Jaguar)
- the Eastern Himalayan Corridor
- China-Russia Tiger Corridor
- Tandai Tiger Corridor

- the European Green Belt

- The Siju-Rewak Corridor, located in the Garo Hills of India, protects an important population of elephants(thought to be approximately 20% of all the elephants that survive in the country).This corridor project links together the Siju Wildlife Sanctuary and the Rewak Reserve Forest in Meghalaya State, close to the India-Bangladesh border. This area lies within the meeting place of the Himalayan Mountain Range and the Indian Peninsula and contains at least 139 other species of mammal, including tiger, clouded leopard and the Himalayan black bear.

- the Ecologische Hoofdstructuur is a network of corridors and habitats created for wildlife in the Netherlands.

Evaluation

Some animal species are much more apt to use habitat corridors than others depending on what their migration and mating patterns are like. For example, many cases of birds and butterflies successfully using corridors have been observed. Less successful stories have come out of mammals such as deer. How effective a corridor is may simply rely on what species it is directed towards (Tewskbury 2002). Corridors created with birds in mind may be more successful because they are highly migratory to begin with.

Human interference is almost inevitable with the quickly increasing population. The goal behind habitat corridors shows the most hope for solving habitat fragmentation and restoring biodiversity as much as possible. Although there are many positives and negatives, there may be enough positives to continue studying and improving corridors. It is truly difficult to say whether corridors are the solution to increasing biodiversity, because each one must be judged on its own. Each corridor has its own set of standards and goals that may set it apart from another one.

Negatives

A major downfall to habitat corridors is that not much information has been gathered about their success. Due to the lack of positive data, many agencies will not allow corridors to be established because they are unsure of their effectiveness. Another problem with corridors is that they are not as useful as simply preserving land so that it cannot be fragmented. However, it is becoming very difficult to set aside land for nature reserves when road-building, industry, and urban sprawl are all competing for space.

Even if corridors are sought as a solution, it does not necessarily mean that animals will use them. Especially in the case of overpasses, research shows that animals do not like to use them to get to another remnant area of land. Usually overpasses are built over busy highways, and many species are too timid to expose themselves in front of all of the traffic. As more roads and buildings arise, there becomes less space to try to preserve.

Habitat corridors need to be species-specific (not every kind of animal will use every kind of corridor) and corridors can be barriers to some species. For instance plants may use road verges as corridors however some mammals will not cross roads to reach a suitable habitat.

When a corridor is implemented, many times development is so close by, that it becomes difficult to build a wide enough passage. There is usually a very limited amount of space available for corridors, so buffers are not usually added in (Rosenberg 1997). Without a buffer zone, corridors become susceptible to harmful outside factors from city streets, suburb development, rural homes, forestry, cropland, and feedlots.

Unfortunately, another limiting factor to the implementation of corridors is money. With such inconclusive data about the effectiveness of connecting land, it is difficult to get the proper funding. Those who would be in charge of the corridor design and construction would ask such questions as, "What if the corridors affect species negatively?" and "What if they actually aid in the spread of disease and catastrophic events?" Furthermore, there is a possibility that corridors could not only aid in the dispersal of native organisms, but invasive ones, as well (Beier & Loe 1998). If invasive species take over an area they could potentially threaten another species, even to the point of extinction.

Although wildlife corridors have been proposed as solutions to habitat and wildlife population fragmentation, there is little evidence that they are broadly useful as a conservation strategy for all biodiversity in non-developed or less-developed areas, compared to protecting connectivity as the relevant ecological attribute. In other words, corridors may be a useful meme for conservation planning/ers, but the concept has less meaning to wildlife species themselves. Very few wildlife follow easily identified "corridors" or "linkages" (e.g., using computer modeling), instead most species meander and opportunistically move through landscapes during daily, seasonal, and dispersal movement behavior. Wildlife corridors may be useful in highly developed landscapes where they are easily identified as the last remaining and available habitat.

Positives

Habitat corridors may be defenseless against a number of outside influences, but they are still an efficient way of increasing biodiversity. Strips of land aid in the movement of various animal species and pollen and seed dispersal, which is an added benefit to the intended one (M. 2002). For example, when insects carrying pollen or birds carrying seeds travel to another area, plant species effectively get transported, as well.

Another positive aspect of corridors is that they allow both animals and humans to occupy virtually the same areas of land, and thus co-exist where without the corridor this would not be possible. Large animals such as bears can be attracted to residential areas in search of food due to lack of natural resources because of habitat fragmentation. A corridor would provide a passage for the bears to forage in other locations, so that they would not pose as much of a threat to humans.

Wildlife Trade

Wildlife trade refers to the commerce of products that are derived from non-domesticated animals or plants usually extracted from their natural environment or raised under controlled conditions. It can involve the trade of living or dead individuals, tissues such as skins, bones or meat, or other products. Legal wildlife trade is regulated by the United Nations' Convention on International Trade in Endangered Species of Wild Fauna and Flora (CITES), which currently has 170 member countries called *Parties*. Illegal wildlife trade, however, is widespread and constitutes one of the major illegal economic activities, comparable to the traffic of drugs and weapons. Wildlife trade is a serious conservation problem, it has a negative effect on the viability of many wildlife populations and is one of the major threats to the survival of vertebrate species.

Assorted seashells, coral, shark jaws and dried blowfish on sale in Greece.

Framed butterflies, moths, beetles, bats, Emperor scorpions and tarantula spiders on sale in Rhodes, Greece.

Terminology

Wildlife use is a general term for all uses of wildlife products, including ritual or religious uses, consumption of bushmeat and different forms of trade. Wildlife use is usually linked to hunting or poaching. Wildlife trade can be differentiated in legal and illegal trade, and both can have domestic (local or national) or international markets, but they might be often related with each-other.

Wildlife trade often includes the trade of living individuals of wildlife species as companion animals (exotic pet trade) or for zoological institutions. These individuals are sometimes semi-domesticated or bred in captivity for the purpose of trade.

Reasons for Concern

Different forms of wildlife trade or use (utilization, hunting, trapping, collection or over-exploitation) are the second major threat to endangered mammals and it also ranks among the first ten threats to birds, amphibians and cycads.

Wildlife trade threatens the local ecosystem, and puts all species under additional pressure at a time when they are facing threats such as over-fishing, pollution, dredging, deforestation and other forms of habitat destruction. Wildlife is traded alive or dead.

In the food chain, species higher up on the ladder ensure that the species below them do not become too abundant (hence controlling the population of those below them). Animals lower on the ladder are often non-carnivorous (but instead herbivorous) and control the abundance of plant species in a region. Due to the very large amounts of species that are removed from the ecosystem, it is not inconceivable that environmental problems will result, similar to e.g. overfishing, which causes an overabundance of jellyfish.

Survival Rate of Species During Transport

In some instances; such as the sale of chameleons from Madagascar, organisms are transported by boat or via the air to consumers. The survival rate of these is extremely poor (only 1% survival rate). This is undoubtedly caused by the illegal nature; vendors rather not risk that the chameleons were to be discovered and so do not ship them in plain view. Due to the very low survival rate, it also means that far higher amounts of organisms (in this case chameleons) are taken away from the ecosystem, to make up for the losses.

Consequences for Indigenous Peoples

In many instances, tribal people have become the victims of the fallout from poaching. With increased demand in the illegal wildlife trade, tribal people are often direct victims of the measures implemented to protect wildlife. Often reliant upon hunting for food, they are prevented from doing so, and are frequently illegally evicted from their lands following the creation of nature reserves aimed to protect animals. Tribal people are often falsely accused of contributing to the decline of species – in the case of India, for example, they bear the brunt of anti-tiger poaching measures, despite the main reason for the tiger population crash in the 20th century being due to hunting by European colonists and Indian elites. In fact, contrary to popular belief, there is strong evidence to show that they effectively regulate and manage animal populations.

Illegal Wildlife Trade

Shark fin for sale in Hong Kong

Interpol has estimated the extent of the illegal wildlife trade between $10 billion and $20 billion per year. While the trade is a global one, with routes extending to every continent, conservationists say the problem is most acute in Southeast Asia. There, trade linkages to key markets in China, the United States, and the European Union; lax law enforcement; weak border controls; and the perception of high profit and low risk contribute to large-scale commercial wildlife trafficking. The ASEAN Wildlife Enforcement Network (ASEAN-WEN) ASEAN Wildlife Enforcement Network, supported by the U.S. Agency for International Development and external funders, is one response to the region's illegal wildlife trade networks.

Asia and Africa

Notable trade hubs of the wildlife trade include Suvarnabhumi International Airport in Bangkok, which offers smugglers direct jet service to Europe, the Middle East, North America and Africa. The Chatuchak weekend market in Bangkok is a known center of illicit wildlife trade, and the sale of lizards, primates, and other endangered species has been widely documented. Trade routes connecting in Southeast Asia link Madagascar to the United States (for the sale of turtles, lemurs, and other primates), Cambodia to Japan (for the sale of slow lorises as pets), and the sale of many species to China.

Morocco has been identified as a transit country for wildlife moving from Africa to Europe due to its porous borders with Spain. Wildlife is present in the markets as photo props, sold for decoration, used in medicinal practices, sold as pets and used to decorate shops. Large numbers of reptiles are sold in the markets, especially spur-thighed tortoises. Although leopards have most likely been extirpated from Morocco, their skins can regularly be seen sold openly as medicinal products or decoration in the markets.

Despite international and local laws designed to crack down on the trade, live animals and animal parts — often those of endangered or threatened species - are sold in open-air markets throughout Asia. The animals involved in the trade end up as trophies, or in specialty restaurants. Some are used in traditional Chinese medicine (TCM). Despite the name, elements of TCM are widely adopted throughout East and Southeast Asia, among both Chinese and non-Chinese communities.

The trade also includes demand for exotic pets, and consumption of wildlife for meat. Large volumes of fresh water tortoises and turtles, snakes, pangolins and monitor lizards are consumed as meat in Asia, including in specialty restaurants that feature wildlife as gourmet dining.

In Thailand the Tiger Temple was closed in 2016 accused of clandestine exchange of tigers.

South America

Although the volume of animals traded may be greater in Southeast Asia, animal trading in Latin America is widespread as well.

In open air Amazon markets in Iquitos and Manaus, a variety of rainforest animals are sold openly as meat, such as agoutis, peccaries, turtles, turtle eggs, walking catfish, etc. In addition, many species are sold as pets. The keeping of parrots and monkeys as pets by villagers along the Amazon is commonplace. But the sale of these "companion" animals in open markets is rampant. Capturing the baby tamarins, marmosets, spider monkeys, saki monkeys, etc., in order to sell them, often requires shooting the mother primate out of a treetop with her clinging child; the youngster may or may not survive the fall. With the human population increasing, such practices have a serious impact on the future prospects for many threatened species. The United States is a popular destination for Amazonian rainforest animals. They are smuggled across borders the same way illegal drugs are - in the trunks of cars, in suitcases, in crates disguised as something else.

In Venezuela more than 400 animal species are involved in subsistence hunting, domestic and international (illegal) trade. These activities are widespread and might overlap in many regions, although they are driven by different markets and target different species.

Online

Through both deep web (password protected, encrypted) and dark web (special portal browsers) markets, participants can trade and transact illegal substances, including wildlife. However the amount of activity is still negligible compared to the amount on the open or surface web. As stated in an examination of search engine key words relating to wildlife trade in an article published by *Conservation Biology*, "This negligible level of activity related to the illegal trade of wildlife on the dark web relative to the open and increasing trade on the surface web may indicate a lack of successful enforcement against illegal wildlife trade on the surface web."

Legal Wildlife Trade

Legal trade of wildlife has occurred for many species for a number of reasons, including commercial trade, pet trade as well as conservation attempts. Whilst most examples of legal trade of wildlife are as a result of large population numbers or pests, there is potential for the use of legal trade to reduce illegal trade threatening many species. Legalizing the trade of species can allow for more regulated harvesting of animals and prevent illegal over-harvesting.

Many environmentalists, scientists, and zoologists around the world are mostly against legalizing pet trade of invasive or introduced species, as their release into the wild, be it intentional or not, could compete with the indegeneous species, can lead to its endangerment.

Examples of Successful Wildlife Trade

Australia

Crocodiles

Trade of crocodiles in Australia has been largely successful. Saltwater crocodiles (*Crocodylus porosus*) and freshwater crocodiles (*Crocodylus johnstoni*) are listed under CITES Appendix II. Commercial harvesting of these crocodiles occurs in Northern Territory, Queensland and Western Australia, including harvesting from wild populations as well as approved captive breeding programs based on quotas set by the Australian government.

Kangaroos

Kangaroos are currently legally harvested for commercial trade and export in Australia. There are a number of species included in the trade including:

- Red kangaroo (*Macropus rufus*)
- Eastern grey kangaroo (*M. giganteus*)
- Western grey kangaroo (*M.fuliginosus*)
- Common wallaroo (*M. robustus*)

Harvesting of kangaroos for legal trade does not occur in National Parks and is determined by quotas set by state government departments. Active kangaroo management has gained a commercial value in the trade of kangaroo meat, hides and other products.

North America

Alligator

Alligators have been traded commercially in Florida and other American states as part of a management program. The use of legal trade and quotas have allowed management of a species as well as economic incentive for sustaining habitat with greater ecological benefits.

Legalising Trade for Endangered Species

The 15th Conference of the Parties of CITES was held in Doha, Qatar in March 2010.

Under the Convention on International Trade of Endangered Species (CITES), species listed under Appendix I are threatened with extinction, and commercial trade in wild-caught specimens, or products derived from them, is prohibited. This rule applies to all species threatened with extinction, except in exceptional circumstances. Commercial trade of endangered species listed under Appendix II and III is not prohibited, although Parties must provide non-detriment finding to show that the species in the wild is not being unsustainably harvested for the purpose of trade. Specimens of Appendix I species that were bred in captivity for commercial purposes are treated as Appendix II. An example of this is captive-bred saltwater crocodiles, with some wild populations listed in Appendix I and others in Appendix II.

Organizations Addressing Illegal Wildlife Trade

- ASEAN Wildlife Enforcement Network (ASEAN-WEN)
- South Asia Wildlife Enforcement Network (SAWEN)
- Clark R. Bavin National Fish and Wildlife Forensic Laboratory
- FREELAND Foundation
- National Rifle Association
- Species Survival Network
- TRAFFIC, the wildlife trade monitoring network
- Wildlife Alliance

Wildlife Conservation

The Siberian tiger is a subspecies of tiger that is endangered; three subspecies of tiger are already extinct.

Conservation is the practice of protecting wild plant and animal species and their habitats. The goal of wildlife conservation is to ensure that nature will be around for future generations to enjoy and also to recognize the importance of wildlife and wilderness for humans and other species alike. Many nations have government agencies and NGO's dedicated to wildlife conservation, which help to implement policies designed to protect wildlife. Numerous independent non-profit organizations also promote various wildlife conservation causes.

According to the National Wildlife Federation, wildlife in the United States gets a majority of their funding through appropriations from the federal budget, annual federal and state grants, and financial efforts from programs such as the Conservation Reserve Program, Wetlands Reserve Program and Wildlife Habitat Incentives Program. Furthermore, a substantial amount of funding comes from the state through the sale of hunting/fishing licenses, game tags, stamps, and excise taxes from the purchase of hunting equipment and ammunition, which collects around $200 million annually.

Wildlife conservation has become an increasingly important practice due to the negative effects of human activity on wildlife. An endangered species is defined as a population of a living species that is in the danger of becoming extinct because the species has a very low or falling population, or because they are threatened by the varying environmental or prepositional parameters.

Major Dangers to Wildlife

Fewer natural wildlife habitat areas remain each year. Moreover, the habitat that remains has often been degraded to bear little resemblance to the wild areas which existed in the past. Habitat loss due to destruction, fragmentation and degradation of habitat is the primary threat to the survival of wildlife.

- Climate change: Global warming is making hot days hotter, rainfall and flooding heavier, hurricanes stronger and droughts more severe. This intensification of weather and climate extremes will be the most visible impact of global warming in our everyday lives. It is also causing dangerous changes to the landscape of our world, adding stress to wildlife species and their habitat. Since many types of plants and animals have specific habitat requirements, climate change could cause disastrous loss of wildlife species. A slight drop or rise in average rainfall will translate into large seasonal changes. Hibernating mammals, reptiles, amphibians and insects are harmed and disturbed. Plants and wildlife are sensitive to moisture change so, they will be harmed by any change in moisture level. Natural phenomena like floods, earthquakes, volcanoes, lightning, forest fires.

- Unregulated Hunting and poaching: Unregulated hunting and poaching causes a major threat to wildlife. Along with this, mismanagement of forest department and forest guards triggers this problem.

- Pollution: Pollutants released into the environment are ingested by a wide variety of organisms. Pesticides and toxic chemical being widely used, making the environment toxic to certain plants, insects, and rodents.

- Perhaps the largest threat is the extreme growing indifference of the public to wildlife, conservation and environmental issues in general. Over-exploitation of resources, i.e., exploitation of wild populations for food has resulted in population crashes (over-fishing and over-grazing for example).

- Over exploitation is the over use of wildlife and plant species by people for food, clothing, pets, medicine, sport and many other purposes. People have always depended on wildlife and plants for food, clothing, medicine, shelter and many other needs. But today we are taking more than the natural world can supply. The danger is that if we take too many individuals of a species from their natural environment, the species may no longer be able to survive. The loss of one species can affect many other species in an ecosystem. The hunting, trapping, collecting and fishing of wildlife at unsustainable levels is not something new. The passenger pigeon was hunted to extinction, early in the last century, and over-hunting nearly caused the extinction of the American bison and several species of whales.

- Deforestation: Humans are continually expanding and developing, leading to an invasion of wildlife habitats. As humans continue to grow they clear forested land to crewe more space. This stresses wildlife populations as there are fewer homes and food sources to survive off of.

- Population: The increasing population of human beings is the major threat to wildlife. More people on the globe means more consumption of food, water and fuel, therefore more waste is generated. Major threats to wildlife are directly related to increasing population of human beings. Low population of humans results in less disturbance to wildlife.

Wildlife Conservation as a Government Involvement

In 1972, the Government of India enacted a law called the Wild Life (Protection) Act. In America, the Endangered Species Act of 1973 protects some U.S. species that were in danger from over exploitation, and the Convention on International Trade in Endangered Species of Fauna and Flora (CITES) works to prevent the global trade of wildlife, but there are many species that are not protected from being illegally traded or over-harvested. The World Conservation Strategy was developed in 1980 by the "International Union for Conservation of Nature and Natural Resources" (IUCN) with advice, co-operation and financial assistance of the United Nations Environment Programme (UNEP) and the World Wildlife Fund and in collaboration with the Food and Agriculture Organization of the United Nations (FAO) and the United Nations Educational, Scientific and Cultural Organization (Unesco)" The strategy aims to "provide an intellectual framework and practical guidance for conservation actions." This thorough guidebook covers everything from the intended "users" of the strategy to its very priorities. It even includes a map section containing areas that have large seafood consumption and are therefore endangered by over fishing. The main sections are as follows:

The marking off of a sea turtle nest. Anna Maria, FL. 2012.

- The objectives of conservation and requirements for their achievement:

    1. Maintenance of essential ecological processes and life-support systems.

    2. Preservation of genetic diversity that is flora and fauna.

    3. Sustainable utilization of species and ecosystems.

- Priorities for national action:

    1. A framework for national and sub-national conservation strategies.

    2. Policy making and the integration of conservation and development.

    3. Environmental planning and rational use allocation.

- Priorities for international action:

    1. International action: law and assistance.

    2. Tropical forests and dry lands.

    3. A global programme for the protection of genetic resource areas.

Map sections:

1. Tropical forests

2. Deserts and areas subject to desertification.

Non-government Involvement

As major development agencies became discouraged with the public sector of environmental conservation in the late 1980s, these agencies began to lean their support towards the "private sector" or non-government organizations (NGOs). In a World Bank Discussion Paper it is made apparent that "the explosive emergence of nongovernmental organizations" was widely known to government policy makers. Seeing this rise in NGO support, the U.S. Congress made amendments to the Foreign Assistance Act in 1979 and 1986 "earmarking U.S. Agency for International Development (USAID) funds for biodiversity". From 1990 moving through recent years environmental conservation in the NGO sector has become increasingly more focused on the political and economic impact of USAID given towards the "Environment and Natural Resources". After the terror attacks on the World Trade Centers on September 11, 2001 and the start of former President Bush's War on Terror, maintaining and improving the quality of the environment and natural resources became a "priority" to "prevent international tensions" according to the Legislation on Foreign Relations Through 2002 and section 117 of the 1961 Foreign Assistance Act. Furthermore, in 2002 U.S. Congress modified the section on endangered species of the previously amended Foreign Assistance Act.

Active Non-government Organizations

Many NGOs exist to actively promote, or be involved with wildlife conservation:

*   The Nature Conservancy is a US charitable environmental organization that works to preserve the plants, animals, and natural communities that represent the diversity of life on Earth by protecting the lands and waters they need to survive.

*   World Wide Fund for Nature (WWF) is an international non-governmental organization working on issues regarding the conservation, research and restoration of the environment, formerly named the World Wildlife Fund, which remains its official name in Canada and the United States. It is the world's largest independent conservation organization with over 5 million supporters worldwide, working in more than 90 countries, supporting around 1300 conservation and environmental projects around the world. It is a charity, with approximately 60% of its funding coming from voluntary donations by private individuals. 45% of the fund's income comes from the Netherlands, the United Kingdom and the United States.

*   WildTeam

*   Wildlife Conservation Society

*   Audubon Society

*   Traffic (conservation programme)

*   Born Free Foundation

*   WildEarth Guardians

Bird Conservation

The extinction of the dusky seaside sparrow was caused by habitat loss.

Bird conservation is a field in the science of conservation biology related to threatened birds. Humans have had a profound effect on many bird species. Over one hundred species have gone extinct in historical times, although the most dramatic human-caused extinctions occurred in the Pacific Ocean as humans colonised the islands of Melanesia, Polynesia and Micronesia, during which an estimated 750-1800 species of bird became extinct. According to Worldwatch Institute, many bird populations are currently declining worldwide, with 1,200 species facing extinction in the next century. The biggest cited reason surrounds habitat loss. Other threats include overhunting, accidental mortality due to structural collisions, long-line fishing bycatch, pollution, competition and predation by pet cats, oil spills and pesticide use and climate change. Governments, along with numerous conservation charities, work to protect birds in various ways, including legislation, preserving and restoring bird habitat, and establishing captive populations for reintroductions.

Threats to Birds

Habitat Loss

The most critical threat facing threatened birds is the destruction and fragmentation of habitat. The loss of forests, plains and other natural systems into agriculture, mines, and urban developments, the draining of swamps and other wetlands, and logging reduce potential habitat for many species. In addition the remaining patches of habitat are often too small or fragmented by the construction of roads or other such barriers that cause populations in these fragmented *islands* to become vulnerable to localised extinction. In addition many forest species show limited abilities to disperse and occupy new forest fragments. The loss of tropical rainforest is the most pressing problem, as these forests hold the highest number of species yet are being destroyed quickly. Habitat loss has been implicated in a number of extinctions, including the ivory-billed woodpecker (disputed because of "rediscovery"), Bachman's warbler and the dusky seaside sparrow.

Introduced Species

Arctic foxes introduced to the Aleutian Islands devastated populations of auks; here a least auklet has been taken.

Historically the threat posed by introduced species has probably caused the most extinctions of birds, particularly on islands. most prehistoric human caused extinctions were insular as well. Many island species evolved in the absence of predators and consequently lost many anti-predator behaviours. As humans traveled around the world they brought with them many foreign animals which disturbed these island species. Some of these were unfamiliar predators, like rats, feral cats, and pigs; others were competitors, such as other bird species, or herbivores that degraded breeding habitat. Disease can also play a role; introduced avian malaria is thought to be a primary cause of many extinctions in Hawaii. The dodo is the most famous example of a species that was probably driven to extinction by introduced species (although human hunting also played a role), other species that were victims of introduced species were the Lyall's wren, po'o-uli and the Laysan millerbird. Many species currently threatened with extinction are vulnerable to introduced species, such as the kōkako, black robin, Mariana crow, and the Hawaiian duck.

Hunting and Exploitation

Humans have exploited birds for a very long time, and sometimes this exploitation has resulted in extinction. Overhunting occurred in some instances with a naive species unfamiliar with humans, such as the moa of New Zealand, in other cases it was an industrial level of hunting that led to extinction. The passenger pigeon was once the most numerous species of bird alive (possibly ever), overhunting reduced a species that once numbered in the billions to extinction. Hunting pressure can be for food, sport, feathers, or even come from scientists collecting museum specimens. Collection of great auks for museums pushed the already rare species to extinction.

The harvesting of parrots for the pet trade has led to many species becoming endangered. Between 1986 and 1988 two million parrots were legally imported into the US alone. Parrots are also illegally smuggled between countries, and rarer species can command high prices.

Hybridisation

Hybridisation may also endanger birds, damaging the gene stock. For example, the American black duck has been often reported hybridising with the mallard, starting a slow decline.

Gamebird hybrids are particularly common and many breeders produce hybrids that may be accidentally or intentionally introduced into the wild.

Other Threats

Birds face a number of other threats. Pollution has led to serious declines in some species. The pesticide DDT was responsible for thinning egg shells in nesting birds, particularly seabirds and birds of prey that are high on the food chain. Seabirds are also vulnerable to oil spills, which destroy the plumage's waterproofing, causing the birds to drown or die of hypothermia. Light pollution can also have a damaging effect on some species, particularly nocturnal seabirds such as petrels.

Seabirds face another threat in the form of bycatch, where birds in the water become tangled in fishing nets or hooked on lines set out by long-line fisheries. As many as 100,000 albatrosses are hooked and drown each year on tuna lines set out by long-line fisheries.

Birds are also threatened by high rise buildings, communications towers, and wind farms; estimates vary from about 3.5 to 975 million birds a year in the North America alone. The largest source of human-related bird death is due to glass windows, which kill 100–900 million birds a year. The next largest sources of human-caused death are hunting (100+ million), house cats (100 million), cars and trucks (50 to 100 million), electric power lines (174 million), and pesticides (67 million). Birds are also killed in large quantities by flying into communication tower guidelines, usually after being attracted by tower lights. This phenomenon is called towerkill and is responsible for 5–50 million birds deaths a year. The presence of towers may seriously impact endangered species living in the vicinity. Birds can also be killed by heat while flying above solar power plants: in 2015, biologists working for the state of California estimated that 3,500 birds died at a single solar plant in the span of a year, "many of them burned alive while flying through a part of the solar installment where air temperatures can reach 1,000 degrees Fahrenheit."

Conservation Techniques

Scientists and conservation professionals have developed a number of techniques to protect bird species. These techniques have had varying levels of success.

Captive Breeding

Captive breeding, or *ex-situ* conservation, has been used in a number of instances to save species from extinction. The principle is to create a viable population of a species in either zoos or breeding facilities, for later reintroduction back into to the wild. As such a captive population can either serve as an insurance against the species going extinct in the wild or as a last-ditch effort in situations where conservation in the wild is impossible. Captive breeding has been used to save several species from extinction, the most famous example being the California condor, a species that declined to less than thirty birds. In order to save the California condor the decision was made to take every individual left in the wild into captivity. From these 22 individuals a breeding programme began that brought the numbers up to 273 by 2005. An even more impressive recovery was that of the Mauritius kestrel, which by 1974 had dropped to only four individuals, yet by 2006 the population was 800.

Reintroduction and Translocations

Reintroductions of captive bred populations can occur to replenish wild populations of an endangered species, to create new populations or to restore a species after it has become extinct in the wild. Reintroductions helped bring the wild populations of Hawaiian geese (nene) from 30 birds to over 500. The Mauritius kestrel was successfully reintroduced into the wild after its captive breeding programme. Reintroductions can be very difficult and often fail if insufficient preparations are made, as species born in captivity may lack the skills and knowledge needed for life in the wild after living in captivity. Reintroductions can also fail if the causes of a birds decline have not been adequately addressed. Attempts to reintroduce the Bali starling into the wild failed due to continued poaching of reintroduced birds.

The introduction of captives of unknown pedigree can pose a threat to native populations. Domestic fowl have threatened endemic species such as *Gallus g. bankiva* while pheasants such as the ring-necked pheasant and captive cheer pheasants of uncertain origin have escaped into the wild or have been intentionally introduced. Green peafowl of similar mixed origins confiscated from local bird dealers have been released into areas with native wild birds.

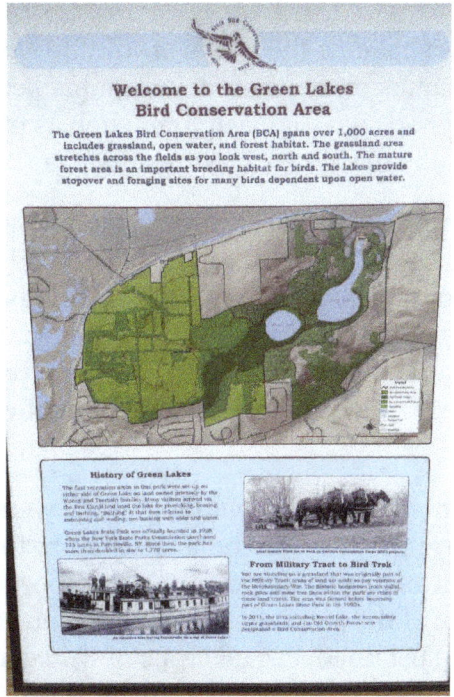

Bird conservation area, Green Lakes State Park, Manlius, New York

Translocations involve moving populations of threatened species into areas of suitable habitat currently unused by the species. There are several reasons for doing this; the creation of secondary populations that act as an insurance against disaster, or in many cases threats faced by the original population in its current location. One famous translocation was of the kakapo of New Zealand. These large flightless parrots were unable to cope with introduced predators in their remaining habitat on Stewart Island, so were moved to smaller offshore islands that had been cleared of predators. From there a recovery programme has managed to maintain and eventually increase their numbers.

Habitat Protection

As the loss and destruction of habitat is the most serious threat facing many bird species, conservation organisations and government agencies tasked with protecting birds work to protect areas of natural habitat. This can be achieved through purchasing land of conservation importance, setting aside land or gazetting it as a national park or other protected area, and passing legislation preventing landowners from undertaking damaging land use practices, or paying them not to undertake those activities. The goals of habitat protection for birds and other threatened animals and plants often conflicts with other stakeholders, such as landowners and businesses, who can face economically damaging restrictions on their activities. Plans to protect crucial habitat for the spotted owl of North America required the protection of large areas of old growth forest in the western United States; this was opposed by logging companies who claimed it would cause job losses and reduced profits.

Wildlife Management

Wildlife management attempts to balance the needs of wildlife with the needs of people using the best available science. Wildlife management can include game keeping, wildlife conservation and pest control. Wildlife management draws on disciplines such as mathematics, chemistry, biology, ecology, climatology and geography to gain the best results.

Wildlife conservation aims to halt the loss in the Earth's biodiversity by taking into consideration ecological principles such as carrying capacity, disturbance and succession and environmental conditions such as physical geography, pedology and hydrology with the aim of balancing the needs of wildlife with the needs of people. Most wildlife biologists are concerned with the preservation and improvement of habitats although rewilding is increasingly being used. Techniques can include reforestation, pest control, nitrification and denitrification, irrigation, coppicing and hedge laying.

Game keeping is the management or control of wildlife for the well being of game and may include killing other animals which share the same niche or predators to maintain a high population of the more profitable species, such as pheasants introduced into woodland. In his 1933 book *Game Management*, Aldo Leopold, one of the pioneers of wildlife management as a science, defined it as "the art of making land produce sustained annual crops of wild game for recreational use".

Pest control is the control of real or perceived pests and can be used for the benefit of wildlife, farmers, game keepers or safety reasons. In the United States, wildlife management practices are often implemented by a governmental agency to uphold a law, such as the Endangered Species Act.

In the United Kingdom, wildlife management undertaken by several organizations including government bodies such as the Forestry Commission, Charities such as the RSPB and The Wildlife Trusts and privately hired gamekeepers and contractors. Legislation has also been passed to protect wildlife such as the Wildlife and Countryside Act 1981. The UK government also give farmers subsidies through the Countryside Stewardship Scheme to improve the conservation value of their farms.

History

Game Laws

The Game Act 1831 protects game birds in England and Wales

The history of wildlife management begins with the game laws, which regulated the right to kill certain kinds of fish and wild animal (game). In Britain game laws developed out of the forest laws, which in the time of the Norman kings were very oppressive. Under William the Conqueror, it was as great a crime to kill one of the king's deer as to kill one of his subjects. A certain rank and standing, or the possession of a certain amount of property, were for a long time qualifications indispensably necessary to confer upon any one the right of pursuing and killing game.

The Game Act of 1831 protected game birds by establishing close seasons when they could not be legally taken. The act made it lawful to take game only with the provision of a game licence and provided for the appointment of gamekeepers around the country. The purposes of the law was to balance the needs for preservation and harvest and to manage both environment and populations of fish and game.

Early game laws were also enacted in the US; - in 1839 Rhode Island closed the hunting season for white-tailed deer from May to November. Other regulations during this time focused primarily on restricting hunting. At this time, lawmakers did not consider population sizes or the need for preservation or restoration of wildlife habitats.

Emergence of Wildlife Conservation

The late 19th century saw the passage of the first pieces of wildlife conservation legislation and the establishment of the first nature conservation societies. The Sea Birds Preservation Act of 1869 was passed in Britain as the first nature protection law in the world after extensive lobbying from the Association for the Protection of Seabirds.

The Royal Society for the Protection of Birds was founded as the Plumage League in 1889 by Emily Williamson at her house in Manchester as a protest group campaigning against the use of great crested grebe and kittiwake skins and feathers in fur clothing. The group gained popularity and eventually amalgamated with the Fur and Feather League in Croydon to form the RSPB. The Society attracted growing support from the suburban middle-classes as well as support from many other influential figures, such as the ornithologist Professor Alfred Newton.

The National Trust formed in 1895 with the manifesto to "...promote the permanent preservation, for the benefit of the nation, of lands, ...to preserve (so far practicable) their natural aspect." On 1 May 1899, the Trust purchased two acres of Wicken Fen with a donation from the amateur naturalist Charles Rothschild, establishing the first nature reserve in Britain. Rothschild was a pioneer of wildlife conservation in Britain, and went on to establish many other nature reserves, such as one at Woodwalton Fen, near Huntingdon, in 1910. During his lifetime he built and managed his estate at Ashton Wold in Northamptonshire to maximise its suitability for wildlife, especially butterflies. Concerned about the loss of wildlife habitats, in 1912 he set up the Society for the Promotion of Nature Reserves, the forerunner of The Wildlife Trusts partnership.

During the society's early years, membership tended to be made up of specialist naturalists and its growth was comparatively slow. The first independent Trust was formed in Norfolk in 1926 as the Norfolk Naturalists Trust, followed in 1938 by the Pembrokeshire Bird Protection Society which after several subsequent changes of name is now the Wildlife Trust of South and West Wales and it was not until the 1940s and 1950s that more Naturalists' Trusts were formed in Yorkshire, Lincolnshire, Leicestershire and Cambridgeshire. These early Trusts tended to focus on purchasing land to establish nature reserves in the geographical areas they served.

Wildlife Management in the US

The profession of wildlife management was established in the United States in the 1920s and '30s by Aldo Leopold and others who sought to transcend the purely restrictive policies of the previous generation of conservationists, such as anti-hunting activist William T. Hornaday. Leopold and his close associate Herbert Stoddard, who had both been trained in scientific forestry, argued that modern science and technology could be used to restore and improve wildlife habitat and thus produce abundant "crops" of ducks, deer, and other valued wild animals.

The institutional foundations of the profession of wildlife management were established in the 1930s, when Leopold was granted the first university professorship in wildlife management (1933, University of Wisconsin, Madison), when Leopold's textbook 'Game Management' was published (1933), when The Wildlife Society was founded, when the Journal of Wildlife Management began publishing, and when the first Cooperative Wildlife Research Units were established. Conservationists planned many projects throughout the 1940s. Some of which included the harvesting of female mammals such as deer to decrease rising populations. Others included waterfowl and wetland research. The Fish and Wildlife Management Act was put in place to urge farmers to plant food for wildlife and to provide cover for them.

In 1937, the Federal Aid in Wildlife Restoration Act (also known as the Pittman-Robertson Act) was passed in the U.S.. This law was an important advancement in the field of wildlife management. It placed a 10% tax on sales of guns and ammunition. The funds generated were then distributed to the states for use in wildlife management activities and research. This law is still in effect today.

Wildlife management grew after World War II with the help of the GI Bill and a postwar boom in recreational hunting. An important step in wildlife management in the United States national parks occurred after several years of public controversy regarding the forced reduction of the elk population in Yellowstone National Park. In 1963, United States Secretary of the Interior Stewart Udall appointed an advisory board to collect scientific data to inform future wildlife management.

In a paper known as the Leopold Report, the committee observed that culling programs at other national parks had been ineffective, and recommended active management of Yellowstone's elk population.

Elk overpopulation in Yellowstone is thought by many wildlife biologists, such as Douglas Smith, to have been primarily caused by the extirpation of wolves from the park and surrounding environs. After wolves were removed, elk herds increased in population, reaching new highs during the mid-1930s. The increased number of elk apparently resulted in overgrazing in parts of Yellowstone. Park officials decided that the elk herd should be managed. For approximately thirty years, the park elk herds were culled: Each year some were captured and shipped to other locations, a certain number were killed by park rangers, and hunters were allowed to take more elk that migrated outside the park. By the late 1960s the herd populations dropped to historic lows (less than 4,000 for the Northern Range herd). This caused outrage among both conservationists and hunters. The park service stopped culling elk in 1968. The elk population then rebounded. Twenty years later there were 19,000 elk in the Northern Range herd, a historic high.

Since the tumultuous 1970s, when animal rights activists and environmentalists began to challenge some aspects of wildlife management, the profession has been overshadowed by the rise of conservation biology. Although wildlife managers remain central to the implementation of the Endangered Species Act and other wildlife conservation policies, conservation biologists have shifted the focus of conservation away from wildlife management's concern with the protection and restoration of single species and toward the maintenance of ecosystems and biodiversity.

Types of Wildlife Management

There are two general types of wildlife management:

- Manipulative management acts on a population, either changing its numbers by direct means or influencing numbers by the indirect means of altering food supply, habitat, density of predators, or prevalence of disease. This is appropriate when a population is to be harvested, or when it slides to an unacceptably low density or increases to an unacceptably high level. Such densities are inevitably the subjective view of the land owner, and may be disputed by animal welfare interests.

- Custodial management is preventive or protective. The aim is to minimize external influences on the population and its habitat. It is appropriate in a national park where one of the stated goals is to protect ecological processes. It is also appropriate for conservation of a threatened species where the threat is of external origin rather than being intrinsic to the system. Feeding of animals by visitors is generally discouraged.

Opposition

The control of wildlife through killing and hunting has been criticized by animal rights and animal welfare activists. Critics object to the real or perceived cruelty involved in some forms of wildlife management.

Environmentalists have also opposed hunting where they believe it is unnecessary or will negatively affect biodiversity. Critics of game keeping note that habitat manipulation and predator control are often used to maintain artificially inflated populations of valuable game animals (including introduced exotics) without regard to the ecological integrity of the habitat.

Game keepers in the UK claim it to be necessary for wildlife conservation as the amount of countryside they look after exceeds by a factor of nine the amount in nature reserves and national parks.

Management of Hunting Seasons

Wildlife management studies, research and lobbying by interest groups help designate times of the year when certain wildlife species can be legally hunted, allowing for surplus animals to be removed. In the United States, hunting season and bag limits are determined by guidelines set by the United States Fish and Wildlife Service for migratory game such as waterfowl and other migratory gamebirds. The hunting season and bag limits for state regulated game species such as deer are usually determined by State game Commissions, which are made up of representatives from various interest groups, wildlife biologists, and researchers.

Open and closed season on deer in the UK is legislated for in the Deer Act 1991 and the Deer Act (Scotland) 1996.

Open Season

Open season is when wildlife is allowed to be hunted by law and is usually not during the breeding season. Hunters may be restricted by sex, age or class of animal, for instance there may be an open season for any male deer with 4 points or better on at least one side.

Limited Entry

Where the number of animals taken is to be tightly controlled, managers may have a type of lottery system called limited. Many apply, few are chosen. These hunts may still have age, sex or class restrictions.

Closed Season

Closed season is when wildlife is protected from hunting and is usually during its breeding season. Closed season is enforced by law, any hunting during closed season is punishable by law and termed as illegal hunting or poaching.

Type of Weapon Used

In wildlife management one of the conservation principles is that the weapon used for hunting should be the one that causes the least damage to the animal and is sufficiently effective so that it hits the target. Given State and Local laws, types of weapon can also vary depending on type, size, sex of game and also the geographical layout of that specific hunting area.

Wildlife Crossing

Wildlife crossings are structures that allow animals to cross human-made barriers safely. Wildlife crossings may include: underpass tunnels, viaducts, and overpasses (mainly for large or herd-type animals); amphibian tunnels; fish ladders; Canopy bridge (especially for monkeys and squirrels), tunnels and culverts (for small mammals such as otters, hedgehogs, and badgers); green roofs (for butterflies and birds).

Florida State Route 46 was elevated over this underpass. Notice the channeling f
ences on either side of the crossing.

Wildlife crossings are a practice in habitat conservation, allowing connections or reconnections between habitats, combating habitat fragmentation. They also assist in avoiding collisions between vehicles and animals, which in addition to killing or injuring wildlife may cause injury to humans and property damage.

Similar structures can be used for domesticated animals, such as cattle creeps.

Roads and Habitat Fragmentation

Camel Crossing in Kuwait

Habitat fragmentation occurs when human-made barriers such as roads, railroads, canals, electric power lines, and pipelines penetrate and divide wildlife habitat (Primack 2006). Of these, roads have the most widespread and detrimental impacts (Spellerberg 1998). Scientists estimate that the system of roads in the United States impacts the ecology of at least one-fifth of the land area of the country (Forman 2000). For many years ecologists and conservationists have documented the adverse relationship between roads and wildlife. Jaeger et al. (2005) identify four ways that roads and traffic detrimentally impact wildlife populations: (1) they decrease habitat amount and quality, (2) they increase mortality due to wildlife-vehicle collisions (road kill), (3) they prevent access to resources on the other side of the road, and (4) they subdivide wildlife populations into smaller and more vulnerable sub-populations (fragmentation). Habitat fragmentation can lead to extinction or extirpation if a population's gene pool is restricted enough.

The first three impacts (loss of habitat, road kill, and isolation from resources) exert pressure on various animal populations by reducing available resources and directly killing individuals in a population. For instance, Bennett (1991) found that road kills do not pose a significant threat to healthy populations but can be devastating to small, shrinking, or threatened populations. Road mortality has significantly impacted a number of prominent species in the United States, including white-tailed deer (*Odocoileus virginianus*), Florida panthers (*Puma concolor coryi*), and black bears (*Ursus americanus*) (Clevenger et al. 2001). In addition, habitat loss can be direct, if habitat is destroyed to make room for a road, or indirect, if habitat quality close to roads is compromised due to emissions from the roads (e.g. noise, light, runoff, pollution, etc.) (Jaeger et al. 2005). Finally, species that are unable to migrate across roads to reach resources such as food, shelter and mates will experience reduced reproductive and survival rates, which can compromise population viability (Noss et al., 1996).

In addition to the first three factors, numerous studies have shown that the construction and use of roads is a direct source of habitat fragmentation (Spellerberg 1998). As mentioned above, populations surrounded by roads are less likely to receive immigrants from other habitats and as a result, they suffer from a lack of genetic diversity. These small populations are particularly vulnerable to extinction due to demographic, genetic, and environmental stochasticity because they do not contain enough alleles to adapt to new selective pressures such as changes in temperature, habitat, and food availability (Primack 2006).

The relationship between roads and habitat fragmentation is well documented. One study found that roads contribute more to fragmentation in forest habitats than clear cuts (Reed et al. 1996). Another study concluded that road fragmentation of formerly contiguous forest in eastern North America is the primary cause for the decline of forest bird species and has also significantly harmed small mammals, insects, and reptiles in the United States (Spellerberg 1998). After years of research, biologists agree that roads and traffic lead to habitat fragmentation, isolation and road kill, all of which combine to significantly compromise the viability of wildlife populations throughout the world.

Wildlife-vehicle Collisions

In addition to conservation concerns, wildlife-vehicle collisions have a significant cost for human populations because collisions damage property and injure and kill passengers and drivers. Bruinderink & Hazebroek (1996) estimated the number of collisions with ungulates in traffic in Europe at 507,000 per year, resulting in 300 people killed, 30,000 injured, and property damage exceeding $1 billion. In parallel, 1.5 million traffic accidents involving deer in the United States cause an estimated $1.1 billion in vehicle damage each year (Donaldson 2005).

The conservation issues associated with roads (wildlife mortality and habitat fragmentation) coupled with the substantial human and economic costs resulting from wildlife-vehicle collisions have caused scientists, engineers, and transportation authorities to consider a number of mitigation tools for reducing the conflict between roads and wildlife. Of the currently available options, structures known as wildlife crossings have been the most successful at reducing both habitat fragmentation and wildlife-vehicle collisions caused by roads (Knapp et al. 2004, Clevenger, 2006).

"Animals' Bridge," on the Flathead Indian Reservation in Montana, used by grizzly and black bears, deer, elk, mountain lions, and others.

Wildlife crossings are structural passages beneath or above roadways that are designed to facilitate safe wildlife movement across roadways (Donaldson 2005). In recent years, conservation biologists and wildlife managers have advocated wildlife crossings coupled with roadside fencing as a way to increase road permeability and habitat connectivity while decreasing wildlife-vehicle collisions. Wildlife crossing is the umbrella term encompassing underpasses, overpasses, ecoducts, green bridges, amphibian/small mammal tunnels, and wildlife viaducts (Bank et al. 2002). All of these structures are designed to provide semi-natural corridors above and below roads so that animals can safely cross without endangering themselves and motorists.

History and Location

Written reports of rough fish ladders date to 17th-century France, where bundles of branches were used to create steps in steep channels to bypass obstructions. A version was patented in 1837 by Richard McFarlan of Bathurst, New Brunswick, Canada, who designed a fishway to bypass a dam at his water-powered lumber mill. In 1880, the first fish ladder was built in Rhode Island, United States, on the Pawtuxet Falls Dam. As the Industrial Age advanced, dams and other river obstructions became larger and more common, leading to the need for effective fish by-passes.

The first overland wildlife crossings were constructed in France during the 1950s (Chilson 2003). European countries including the Netherlands, Switzerland, Germany, and France have been using various crossing structures to reduce the conflict between wildlife and roads for several decades and use a variety of overpasses and underpasses to protect and re-establish wildlife such as: amphibians, badgers, ungulates, invertebrates, and other small mammals (Bank et al. 2002).

The Humane Society of the United States reports that the more than 600 tunnels installed under major and minor roads in the Netherlands have helped to substantially increase population levels of the endangered European badger. The longest "ecoduct" viaduct, near Crailo in the Netherlands, runs 800 m and spans a highway, railway and golf course.

A Terrapin crossing sign and a highway barrier designed for crossing at the end of the
F.J. Torras causeway at St. Simons Island, Georgia, USA (2015)

Wildlife crossings are becoming increasingly common in Canada and the United States. Recognizable wildlife crossings are found in Banff National Park in Alberta, where vegetated overpasses provide safe passage over the Trans-Canada Highway for bears, moose, deer, wolves, elk, and many other species (Clevenger 2007). The 24 wildlife crossings in Banff were constructed as part of a road improvement project in 1978 (Clevenger 2007). In the United States, thousands of wildlife crossings have been built in the past 30 years, including culverts, bridges, and overpasses. These have been used to protect mountain goats in Montana, spotted salamanders in Massachusetts, bighorn sheep in Colorado, desert tortoises in California, and endangered Florida panthers in Florida (Chilson 2003).

The first wildlife crossing in the Canadian province of Ontario was built in 2010, along Ontario Highway 69 between Sudbury and Killarney, as part of the route's ongoing freeway conversion.

Costs and Benefits

The benefits derived from constructing wildlife crossings to extend wildlife migration corridors over and under major roads appear to outweigh the costs of construction and maintenance. One study estimates that adding wildlife crossings to a road project is a 7-8% increase in the total cost of the project (Bank et al. 2002). Theoretically, the monetary costs associated with constructing and maintaining wildlife crossings in ecologically important areas are trumped by the benefits associated with protecting wildlife populations, reducing property damage to vehicles, and saving the lives of drivers and passengers by reducing the number of collisions caused by wildlife.

A study completed for the Virginia Department of Transportation estimated that underpasses for wildlife become cost effective, in terms of property damage, when they prevent between 2.6 and 9.2 deer-vehicle collisions per year, depending on the cost of the underpass. Approximately 300 deer crossed through the underpasses in the year the study took place (Donaldson 2005).

Effectiveness

A number of studies have been conducted to determine the effectiveness of wildlife corridors at providing habitat connectivity (by providing viable migration corridors) and reducing wildlife-vehicle collisions. The effectiveness of these structures appears to be highly site-specific (due to differences in location, structure, species, habitat, etc.) but crossings have been beneficial to a number of species in a variety of locations. Some of the wildlife crossing success stories are detailed below.

Banff National Park

Banff National Park offers one of the best opportunities to study the effectiveness of wildlife crossings because the park contains a wide variety of species and is bisected by a large commercial road called the Trans-Canada Highway (TCH). To reduce the effects of the four-lane TCH, 24 wildlife crossings (22 underpasses and two overpasses) were built to ensure habitat connectivity and protect motorists (Clevenger 2007). In 1996, Parks Canada developed a contract with university researchers to assess the effectiveness of the crossings. The past decade has produced a number of publications that analyze the crossings' impact on various species and overall wildlife mortality.

Wildlife overpass in Banff National Park, Canada

Using a variety of techniques to monitor the crossings over the last 25 years, scientists report that 10 species of large mammals (including deer, elk, black bear, grizzly bear, mountain lion, wolf, moose, and coyote) have used the 24 crossings in Banff a total of 84,000 times as of January 2007 (Clevenger 2007). The research also identified a "learning curve" such that animals need time to acclimate to the structures before they feel comfortable using them. For example, grizzly bear crossings increased from seven in 1996 to more than 100 in 2006, although the actual number of individual bears using the structures remained constant over this time at between 2 and 4 bears (Parks Canada, unpublished results). A similar set of observations was made for wolves, with crossings increasing from two to approximately 140 over the same 10-year period. However, in this case the actual number of wolves in the packs using the crossings increased dramatically, from a low of two up to a high of over 20 individuals. In continuation with these positive results, Clevenger et al. (2001) reported that the use of wildlife crossings and fencing reduced traffic-induced mortality of large ungulates on the TCH by more than 80 percent. Recent analysis for carnivores showed results were not as positive however, with bear mortality increasing by an average of 116 percent in direct parallel to an equal doubling of traffic volumes on the highway, clearly showing no effect of fencing to reduce bear mortality (Hallstrom, Clevenger, Maher and Whittington, in prep). Research on the crossings in Banff has thus shown mixed value of wildlife crossings depending on the species in question.

Parks Canada is currently planning to build 17 additional crossing structures across the TCH to increase driver safety near the hamlet of Lake Louise. Lack of effectiveness of standard fencing in reducing bear mortality demonstrates that additional measures such as wire 'T-caps' on the fence may be needed for fencing to mitigate effectively for bears (Hallstrom, Clevenger, Maher and Whittington, in prep).

Collier and Lee Counties in Florida

Twenty-four wildlife crossings (highway underpasses) and 12 bridges modified for wildlife have been constructed along a 40-mile stretch of Interstate 75 in Collier and Lee counties in Florida (Scott 2007). These crossings are specifically designed to target and protect the endangered Florida panther, a subspecies of mountain lion found in the southeastern United States. Scientists estimate that there are only 80-100 Florida panthers alive in the wild, making them one of the most endangered large mammals in North America (Foster and Humphrey, 1995). The Florida panther is particularly vulnerable to wildlife-vehicle collisions, which claimed 11 panthers in 2006 and 14 panthers in 2007 (Scott 2007).

The Florida Fish and Wildlife Conservation Commission (FWC) has used a number of mitigation tools in an effort to protect Florida panthers and the combination of wildlife crossings and fences have proven the most effective (Scott 2007). As of 2007, no panthers have been killed in areas equipped with continuous fencing and wildlife crossings and the FWC is planning to construct many more crossing structures in the future. The underpasses on I-75 also appeared to benefit bobcats, deer, and raccoons and significantly reduced wildlife-vehicle collisions along the interstate (Foster and Humphrey, 1995).

Underpasses in Southern California

Wildlife crossings have also been important for protecting biodiversity in several areas of southern California. In San Bernardino County, biologists have erected fences along State Route 58 to complement underpasses (culverts) that are being used by the threatened desert tortoise. Tortoise deaths on the highway declined by 93% during the first four years after the introduction of the fences, proving that even makeshift wildlife crossings (storm-drainage culverts in this case) have the ability to increase highway permeability and protect sensitive species (Chilson 2003). Additionally, studies by Haas (2000) and Lyren (2001) report that underpasses in Orange, Riverside, and Los Angeles Counties have drawn significant use from a variety of species including bobcats, coyotes, gray fox, mule deer, and long-tailed weasels. These results could be extremely important for wildlife conservation efforts in the region's Puente Hills and Chino Hills links, which have been increasingly fragmented by road construction (Haas 2000).

Ecoducts, Netherlands

One of the two wildlife crossings spanning the A50 highway on the Veluwe in the Netherlands.

The Netherlands has over 66 wildlife crossings (overpasses and ecoducts) that have been used to protect the endangered European badger, as well as populations of wild boar, red deer, and roe deer. As of 2012, the Veluwe, 1000 square kilometers of woods, heathland and drifting sands, the largest lowland nature area in North Western Europe, contains nine ecoducts, 50 meters wide on average, that are used to shuttle wildlife across highways that transect the Veluwe. The first two ecoducts on the Veluwe were built in 1988 across the A50 when the highway was constructed. Five of the other ecoducts on the Veluwe were built across existing highways, one was built across a two lane provincial road. The two ecoducts across the A50 were used by nearly 5,000 deer and wild boar during a one-year period (Bank et al. 2002). The Netherlands also boasts the world's longest ecoduct-wildlife overpass called the Natuurbrug Zanderij Crailoo (sand quarry nature bridge at Crailo) (Danby 2004). The massive structure, completed in 2006, is 50 m wide and over 800 m long and spans a railway line, business park, river, roadway, and sports complex (Danby 2004). Monitoring is currently underway to examine the effectiveness of this innovative project combining wildlife protection with urban development. The oldest wildlife passage is Zeist West - A 28, opened in 1988.

Slaty Creek Wildlife Underpass, Calder Freeway, Black Forest, Australia

Another case study of the effectiveness of wildlife crossings comes from an underpass built to minimize the ecological impact of the Calder Freeway as it travels through the Black Forest in Victoria, Australia. In 1997, the Victorian Government Roads Corporation built Slaty Creek wildlife underpass at a cost of $3 million (Abson & Lawrence 2003). Scientists used 14 different techniques to monitor the underpass for 12 months in order to determine the abundance and diversity of species using the underpass (Abson & Lawrence 2003). During the 12-month period, 79 species of fauna were detected in the underpass (compared with 116 species detected in the surrounding forest) including amphibians, bats, birds, koalas, wombats, gliders, reptiles, and kangaroos (Abson & Lawrence 2003). The results indicate that the underpass could be useful to a wide array of species but the authors suggest that Slaty Creek could be improved by enhanced design and maintenance of fencing to minimise road kill along the Calder Freeway and by attempting to exclude introduced predators such as cats and foxes from the area.

The ARC International Wildlife Crossing Infrastructure Design Competition

In 2010, ARC Solutions - an interdisciplinary partnership - initiated the International Wildlife Crossing Infrastructure Design Competition for a wildlife crossing over Interstate 70 near Denver, Colorado., and designers had to account for many challenges unique to the area, including snow and severe weather, high elevation and steep grades, a six-lane roadway, a bike path, and high traffic volumes, as well as multiple species of wildlife, including lynx.

After receiving 36 submissions from nine countries, a jury of internationally acclaimed experts in landscape architecture, engineering, architecture, ecology and transportation selected five finalists in November 2010 to further develop their conceptual designs for a wildlife crossing structure. In January 2011, the team led by HNTB with Michael Van Valkenburgh & Associates (New York) were selected as the winners. The design features a single 100 m (328 ft) concrete span across the highway that is planted with a variety of vegetation types, including a pine-tree forest and meadow grasses, to attract different species to cross. A modular precast concrete design means that much of the bridge can be constructed offsite and moved into place.

Canopy Bridge in Anamalai Tiger Reserve

Many endangered lion-tailed macaques used to get killed while crossing the highway at Pudu-thotam in Valparai, South India. Thanks to the efforts of NGOs and the forest department, several canopy bridges were installed, connecting trees on either side of the road. This helped to lower the numbers of lion-tailed macaques killed in the region. The Environment Conservation Group had initiated a national mission to increase awareness on the importance of adopting roadkill mitigation methods through their mission PATH traveling more than 17,000 kilometers across twenty-two states.

Nuisance Wildlife Management

Deer eating tomato plant

Nuisance wildlife management is the term given to the process of selective removal of problem individuals or populations of specific species of wildlife. Other terms for the field, include wildlife damage management, wildlife control, and animal damage control to name a few. Some species of wildlife may become habituated to man's presence, causing property damage or risking transfer of disease to humans or pets (zoonosis). Many wildlife species coexist with humans very successfully, such as commensal rodents which have become more or less dependent on humans.

Characteristics of Nuisance Species

Deer-damaged tomato plant has been stripped of developing fruit

Typically, species that are most likely to be considered a nuisance by humans have the following characteristics. First, they are adaptable to fragmented habitat. Animals such as Canada geese (*Branta canadensis*) love ponds with low sloping banks leading to lush green grass. Humans love this sort of landscaping too, so it is not surprising that Canada geese have thrived (not to mention the decline in hunting).

Second, these animals are not tied to eating a specific type of food. For example, lynx do not thrive in human impacted environments because they rely so heavily on snowshoe hares. In contrast, raccoons have been very successful in urban landscapes because they can live in attics, chimneys, and even sewers, and can sustain themselves with food gained from trashcans and discarded litter.

Third, successful animals must not pose an obvious significant risk to human health and safety. Animals perceived as grave threats will incur the extreme ire of humans and be under constant threat of humans seeking to eliminate them.

Finally, successful animals in humanized landscapes are often perceived as "cute", at least until they become so numerous that their preferential status becomes diminished. Many wildlife species have the potential of becoming a "nuisance" species, and whether or not a species is regarded as a pest can be directly correlated with the degree to which that animal can be tolerated by humans. For many people, tree squirrels feeding in their yards or gardens are not a problem; a neighbor may feel that these same squirrels nesting in the attic of their house are a nuisance and a fire hazard, due to their habit of gnawing on electrical cables.

Common Nuisance Species

Common wildlife pests include armadillos, skunks, boars, foxes, squirrels, snakes, rats, groundhogs, beavers, opossums, raccoons, bats, moles, deer, mice, coyotes, bears, ravens, seagulls, woodpeckers and pigeons. Some of these species are protected by state or federal regulations, such as bears, ravens, bats, deer, woodpeckers, and coyotes, and a permit may be required to control some species.

Wildlife are usually only pests in certain situations, such as when their numbers become "excessive" in a particular area. Human-induced changes in the environment will often result in increased numbers of a species. For example, piles of scrap building material make excellent sites where rodents can nest. Food left out for household pets is often equally attractive to some wildlife species. In these situations, the wildlife have suitable food and habitat and may become a nuisance.

Controlling Wildlife Damage

The primary objective of any control program should be to reduce damage in a practical, humane and environmentally acceptable manner. Wildlife managers and wildlife control operators (WCOs) use control methods based on the habits and biology of the animals causing damage. By using methods matched to the nuisance species, control efforts will be more effective and will serve to maximize safety to the environment, humans and other animals.

A key to controlling wildlife damage is prompt and accurate determination of which animal is causing the damage. Even someone with no training or experience can sometimes identify the pest by thoroughly examining the damaged area. Because feeding indications of many wildlife species are similar, other signs – such as droppings, tracks, burrows, nests or food caches – are usually needed to make a positive species identification.

Raccoon on a roof

After the wildlife pest is identified, control methods can be chosen appropriate to the animal species involved. Improper control methods may harm but not kill the animal, causing it to become leery of those and other methods in the future. For example, using traps and poison baits improperly or in the wrong situation may teach the animal that the control method is harmful. This may make the animal difficult to control later, even with the correct method.

Four steps lead to a successful nuisance wildlife control program:

- Correctly identify the species causing the problem.

- Alter the habitat, if possible, to make the area less attractive to the wildlife pest.

- Use a control method appropriate to the location, time of year, and other environmental conditions.

- Monitor the site for re-infestation in order to determine if additional control is necessary.

Control Methods

The most commonly used methods for controlling nuisance wildlife around homes and gardens include exclusion, habitat modification, repellents, toxic baits, glue boards, traps and frightening. Wildlife control involves human risks both from possible injury to person and property, but also from zoonotic disease.

Exclusion

Physically excluding an offending animal from the area being damaged or disturbed is often the best and most permanent way to control the problem. Depending upon size of the area to be protected, this control method can range from inexpensive to prohibitively costly.

For example, damage by birds or rabbits to ornamental shrubs or garden plants can be reduced inexpensively by placing bird netting over the plants to keep the pests away. On the other hand, fencing out deer from a lawn or garden can be more costly. Materials needed for exclusion will depend upon the species causing the problem. Large mammals can be excluded with woven wire fences, poly-tape fences, and electric fences; but many communities forbid the use of electric fencing in their jurisdictions. Small mammals and some birds can be excluded with netting, tarp, hardware cloth or any other suitable material; nets come in different weave sizes suitable for different animals to be excluded.

However, exclusion can interfere with the natural movement of wildlife, particularly when exclusion covers large areas of land.

Habitat Modification

Modifying an animal's habitat often provides lasting and cost-effective relief from damage caused by nuisance wildlife. Habitat modification is effective because it limits access to one or more of the requirements for life – food, water or shelter. However, habitat modification, while limiting nuisance wildlife, may also limit desirable species such as songbirds as well.

Rodent- or bat-proofing buildings by sealing cracks and holes prevents these animals from gaining access to suitable habitats where they are not welcome. Storing seed and pet food in tightly closed containers, controlling weeds and garden debris around homes and buildings, and storing firewood and building supplies on racks or pallets above ground level are also practices that can limit or remove the animals' sources of food, water or shelter.

Repellents

Using a repellent that changes the behavior of an animal may lead to a reduction or elimination of damage. Several available repellents, such as objectionable-tasting coatings or odor repellents, may deter wildlife from feeding on plants. Other repellents such as sticky, tacky substances placed on or near windows, trees or buildings may deter many birds and small mammals. Unfortunately, most wildlife soon discover that repellents are not actually harmful, and the animals may quickly become accustomed to the smell, taste or feel of these deterrents.

Chemical repellents applied outdoors will have to be reapplied due to rain or heavy dew, or applied often to new plant growth to be effective. Failure to carefully follow the directions included with repellents can drastically diminish the effectiveness of the product. Some repellents contain toxic chemicals, such as paradichlorobenzene, and are ineffective unless used at hazardous concentrations. Other more natural repellents contain chili pepper or capsaicin extracted from hot peppers.

However, even under the best of conditions, repellents frequently fail to live up to user expectations. The reason for this is twofold. First, many repellents simply don't work. For example, peer-reviewed publications have consistently shown that ultrasonic devices do not drive unwanted animals away. Second, even when the repellent has been shown to work, animals in dire need of food will "hold their nose" and eat anyway because the alternative is essentially death by starvation. Repellents are most successful (referring to products actually demonstrated by peer-reviewed research to be effective) when animals have access to alternative food sources in a different location.

Glue Traps and Boards

Glue traps and boards can be either a lethal or non-lethal method of control. Glue boards can be used to trap small mammals and snakes. Applying vegetable oil will dissolve the glue, allowing for release, but caution must be taken to avoid scratches and bites from the trapped animal.

Live Trapping

Using traps can be very effective in reducing actual population numbers of certain species. However, many species cannot be trapped without a permit. In most cases, homeowners may trap an offending animal within 100 yards of their residence without a permit, however relocation is often illegal.

Traditional live traps such as cage or box traps are easily purchased at most garden centers or hardware stores. These traps allow for safe release of the trapped animal. The release of the animal to another area may be prohibited by state law, or may be regulated by the local Department of Fish and Game. Leghold traps may allow for either release or euthanasia of the trapped animal. Traps such as body-gripping traps, scissor and harpoon traps, as well as rat/mouse snap traps, are nearly always lethal. Knowledge of animal behavior, trapping techniques, and baits is essential for a successful trapping program.(Bornheimer, Shane P. "PreferredWildlifeservices.com" July 2013).

Frightening Devices

Frightening devices such as bells, whistles, horns, clappers, sonic emitters, audio tapes and other sound devices may be quite successful in the short term for repelling an animal from an area. Other objects such as effigies, lights, reflectors and windmills rely on visual stimulation to scare a problem animal away. Often nuisance animals become accustomed to these tactics, and will return later if exposed to these devices daily.

Sonic Nets

In 2013, Dr. John Swaddle and Dr. Mark Hinders at the College of William and Mary created a new method of deterring birds and other animals using benign sounds projected by conventional and directional (parametric) speakers. The initial objectives of the technology were to displace problematic birds from airfields to reduce bird strike risks, minimize agricultural losses due to pest bird foraging, displace nuisance birds that cause extensive repair and chronic clean-up costs, and reduce bird mortality from flying into man-made structures. The sounds, referred to as a "Sonic Net," do not have to be loud and are a combination of wave forms - collectively called "colored" noise - forming non-constructive and constructive interference with how birds and other animals such as deer talk to each other. Technically, the Sonic Nets technology is not a bird or wildlife scarer, but discourages birds and animals from going into or spending time in the target area. The impact to the animals is similar to talking in a crowded room, and since they cannot understand each other they go somewhere else. Early tests at an aviary and initial field trials at a landfill and airfield indicate that the technology is effective and that birds do not habituate to the sound. The provisional and full patents were filed in 2013 and 2014 respectively, and further research and commercialization of the technology are ongoing.

Laws

Before initiating any wildlife control activities, a person must become familiar with applicable federal, state, and local laws. One way to learn these rules is to contact the state's wildlife agency, which is usually responsible for selling hunting and fishing licenses. In general, property owners are permitted to prevent wildlife damage through exclusion and habitat modification, though they may be prohibited from disturbing an occupied nest or den, or directly harming an animal.

Many regulations exist in the United States concerning animal trapping including trap check intervals, usually requiring all traps be checked at least once during a 24-hour period. Some governments permit relocation of wildlife, however humane considerations must be taken into account before relocating wildlife, including population and habitat.

Ethics

There are many ethical considerations in nuisance wildlife management. Some species of wildlife cannot be ethically relocated due to overabundance of competing species, or lack of availability of proper food and habitat. Control during the spring months does run the risk of killing the young by starvation. Proper euthanasia of animals when necessary is also a controversial and sensitive consideration to be taken prior to engaging in nuisance wildlife management, and requires training and certification in some areas of the United States.

Conservation-reliant Species

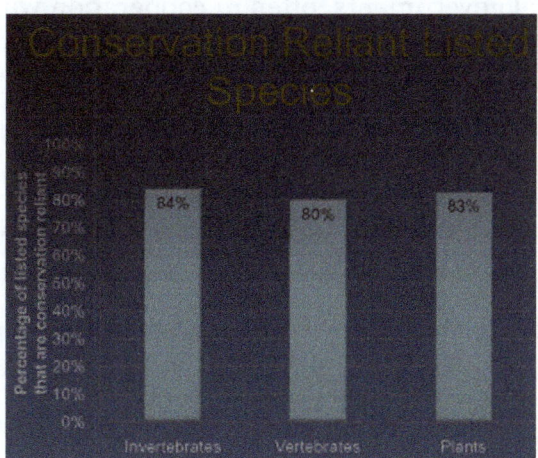

Percentages of United States listed species which are conservation-reliant.

Conservation-reliant species are animal or plant species that require continuing species-specific wildlife management intervention such as predator control, habitat management and parasite control to survive, even when a self-sustainable recovery in population is achieved.

History

The term "conservation-reliant species" grew out of the conservation biology undertaken by *The Endangered Species Act at Thirty Project* (launched 2001) and its popularization by project leader J. Michael Scott. Its first use in a formal publication was in *Frontiers in Ecology and the Environ-*

ment in 2005. Worldwide use of the term has not yet developed and it has not yet appeared in a publication compiled outside North America.

Passages of the 1973 Endangered Species Act (ESA) carried with it the assumption that endangered species would be delisted as their populations recovered. It assumed they would then thrive under existing regulations and the protections afforded under the ESA would no longer be needed. However, eighty percent of species currently listed under the ESA fail to meet that assumption. To survive, they require species-specific conservation interventions (e.g. control of predators, competitors, nest parasites, prescribed burns, altered hydrological processes, etc.) and thus they are conservation-reliant.

Criteria

The criteria for assessing whether a species is conservation-reliant are:

1. Threats to the species' continued existence are known and treatable.

2. The threats are pervasive and recurrent, for example: nest parasites, non-native predators, human disturbance.

3. The threats render the species at risk of extinction, absent ongoing conservation management.

4. Management actions sufficient to counter threats have been identified and can be implemented, for example: prescribed fires, restrictions on grazing or public access, predator or parasite control.

5. National, state or local governments, often in cooperation with private or tribal interests, are capable of carrying out the necessary management actions as long as necessary.

Management Actions

There are five major areas of management action for conservation of vulnerable species:

1. Control of other species may include: control of exotic fauna, exotic flora, other native species and parasites and disease.

2. Control of direct human impacts may include control of grazing, human access, on and off-road vehicles, low impact recreation and illegal collecting and poaching.

3. Pollution control may include control of chemical run-off, siltation, water quality and use of pesticides and herbicides.

4. Active habitat management may include fire management and control, control of soil erosion and waterbodies, habitat restoration and mechanical vegetation control.

5. Artificial population recruitment may include captive propagation (forced immigration) or captive breeding.

Case Study

A prominent example is in India, where tigers, an apex predator and the national animal, are considered a conservation-reliant species. This keystone species can maintain self-sustaining wild populations; however, they require ongoing management actions because threats are pervasive, recurrent and put them at risk of extinction. The origin of these threats are rooted in the changing socio-economic, political and spatial organization of society in India. Tigers have become extinct in some areas because of extrinsic factors such as habitat destruction, poaching, disease, floods, fires and drought, decline of prey species for the same reasons, as well as intrinsic factors such as demographic stochasticity and genetic deterioration.

Bengal tiger at Bannerghatta National Park, Bangalore, India.

Recognizing the conservation reliance of tigers, Project Tiger is establishing a national science-based framework for monitoring tiger population trends in order to manage the species more effectively. India now has 28 tiger reserves, located in 17 states. These reserves cover 37,761 square kilometres (14,580 sq mi) including %1.14 of the total land area of the country. These reserves are kept free of biotic disturbances, forestry operations, collection of minor forest products, grazing and human disturbance. The populations of tigers in these reserves now constitute some of the most important tiger source populations in the country.

Future

The magnitude and pace of human impacts on the environment make it unlikely that substantial progress will be made in delisting many species unless the definition of "recovery" includes some form of active management. Preventing delisted species from again being at risk of extinction may require continuing, species-specific management actions. Viewing "recovery" of "conservation-reliant species" as a continuum of phases rather a simple "recovered/not recovered" status may enhance the ability to manage such species within the framework of the Endangered Species Act. With ongoing loss of habitat, disruption of natural cycles, increasing impacts of non-native invasive species, it is probable that the number of conservation-reliant species will increase.

It has been proposed that development of "recovery management agreements", with legally and biologically defensible contracts would provide for continuing conservation management following delisting. The use of such formalized agreements will facilitate shared management responsibilities between federal wildlife agencies and other federal agencies, and with state, local, and tribal governments, as well as with private entities that have demonstrated the capability to meet the needs of conservation-reliant species.

Conservation-dependent Species

A visualization of the categories in the no-longer used "IUCN 1994 Categories & Criteria (version 2.3)", with *conservation dependent* (LR/cd) highlighted. The category was folded into the "near threatened" category in the 2001 revision, but some species which have not been re-evaluated retain the assessment.

A conservation-dependent species is a species which has been categorised as "Conservation Dependent" ("LR/cd") by the International Union for Conservation of Nature, i.e. as dependent on conservation efforts to prevent it from becoming threatened with extinction. Such species must be the focus of a continuing species-specific and/or habitat-specific conservation programme, the cessation of which would result in the species qualifying for one of the threatened categories within a period of five years.

The category is part of the IUCN 1994 Categories & Criteria (version 2.3), which is no longer used in evaluation of taxa, but persists in the IUCN Red List for taxa evaluated prior to 2001, when version 3.1 was first used. Using the 2001 (v3.1) system these taxa are classed as near threatened, but those that have not been re-evaluated remain with the "Conservation Dependent" category.

As of December 2015, there remains 209 conservation-dependent plant species and 29 conservation-dependent animal species.

Examples of conservation-dependent species include the black caiman, the sinarapan, the salamander mussel, the California ground cricket, and the flowering plant *Garcinia hermonii*.

Conservation-dependent animals

As of December 2015, the IUCN still lists 29 conservation-dependent animal species, and two conservation-dependent subpopulations or stocks.

Mollusks

- *Craspedoma hespericum*
- Bear paw clam
- China clam
- Salamander mussel
- Maxima clam
- Fluted giant clam

Arthropods

- Mono Lake brine shrimp
- Attheyella yemanjae
- Canthocamptus campaneri

- Metacyclops campestris
- Murunducaris juneae
- Muscocyclops bidenatus
- Muscocyclops therasiae
- California ground cricket
- Ponticyclops boscoi
- Spaniacris deserticola
- Stenopelmatus nigrocapitatus
- Thermocyclops parvus

Fish

- Banded mountain zipper loach
- Ceylonese combtail
- Ornate paradisefish
- Sinarapan
- Two spot barb
- Black ruby barb
- Cherry barb
- Pearly rasbora
- Side-striped barb

Reptiles

- Black caiman
- Arrau turtle

Mammals

Subpopulations and stocks

- Bowhead whale (1 subpopulation/stock)
- Northern blue whale (1 subpopulation/stock)

EPBC Act

In Australia, the Environment Protection and Biodiversity Conservation Act 1999 still uses a "Conservation Dependent" category for classifying fauna and flora species. Species recognised as "Conservation Dependent" do not receive special protection, as they are not considered "matters of national environmental significance under the EPBC Act".

The legislation uses categories similar to those of the *IUCN 1994 Categories & Criteria*. It does not, however, have a near threatened category or any other "lower risk" categories.

As of December 2006, only two species have received the status under the act:

- Orange roughy (*Hoplostethus atlanticus*)

- Southern bent-wing bat (*Miniopterus schreibersii bassanii*)

No flora has been given the category under the EPBC Act.

Conservation Biology

Conservation biology is the management of nature and of Earth's biodiversity with the aim of protecting species, their habitats, and ecosystems from excessive rates of extinction and the erosion of biotic interactions. It is an interdisciplinary subject drawing on natural and social sciences, and the practice of natural resource management.

The conservation ethic is based on the findings of conservation biology.

Efforts are made to preserve the natural characteristics of Hopetoun Falls, Australia, without access by visitors being affected.

The term conservation biology and its conception as a new field originated with the convening of "The First International Conference on Research in Conservation Biology" held at the University of California, San Diego in La Jolla, California in 1978 led by American biologists Bruce A. Wilcox and Michael E. Soulé with a group of leading university and zoo researchers and conservationists including Kurt Benirschke, Sir Otto Frankel, Thomas E. Lovejoy, and Jared M. Diamond. The meeting was prompted by the concern over tropical deforestation, disappearing species, eroding genetic diversity within species. The conference and proceedings that resulted sought to initiate the bridging of a gap between theory in ecology and evolutionary genetics on the one hand and conservation policy and practice on the other. Conservation biology and the concept of biological diversity (biodiversity) emerged together, helping crystallize the modern era of conservation science and policy. The inherent multidisciplinary basis for conservation biology has led to new subdisciplines including conservation social science, conservation behavior and conservation physiology. It stimulated further development of conservation genetics which Otto Frankel had originated first but is now often considered a subdiscipline as well.

Description

The rapid decline of established biological systems around the world means that conservation biology is often referred to as a "Discipline with a deadline". Conservation biology is tied closely to ecology in researching the population ecology (dispersal, migration, demographics, effective population size, inbreeding depression, and minimum population viability) of rare or endangered species. Conservation biology is concerned with phenomena that affect the maintenance, loss, and restoration of biodiversity and the science of sustaining evolutionary processes that engender genetic, population, species, and ecosystem diversity. The concern stems from estimates suggesting that up to 50% of all species on the planet will disappear within the next 50 years, which has contributed to poverty, starvation, and will reset the course of evolution on this planet.

Conservation biologists research and educate on the trends and process of biodiversity loss, species extinctions, and the negative effect these are having on our capabilities to sustain the well-being of human society. Conservation biologists work in the field and office, in government, universities, non-profit organizations and industry. The topics of their research are diverse, because this is an interdisciplinary network with professional alliances in the biological as well as social sciences. Those dedicated to the cause and profession advocate for a global response to the current biodiversity crisis based on morals, ethics, and scientific reason. Organizations and citizens are responding to the biodiversity crisis through conservation action plans that direct research, monitoring, and education programs that engage concerns at local through global scales.

History

Natural Resource Conservation

Conscious efforts to conserve and protect *global* biodiversity are a recent phenomenon. Natural resource conservation, however, has a history that extends prior to the age of conservation. Resource ethics grew out of necessity through direct relations with nature. Regulation or communal restraint became necessary to prevent selfish motives from taking more than could be locally sustained, therefore compromising the long-term supply for the rest of the community. This social dilemma with respect to natural resource management is often called the "Tragedy of the Commons".

From this principle, conservation biologists can trace communal resource based ethics throughout cultures as a solution to communal resource conflict. For example, the Alaskan Tlingit peoples and the Haida of the Pacific Northwest had resource boundaries, rules, and restrictions among clans with respect to the fishing of sockeye salmon. These rules were guided by clan elders who knew lifelong details of each river and stream they managed. There are numerous examples in history where cultures have followed rules, rituals, and organized practice with respect to communal natural resource management.

The Mauryan emperor Ashoka around 250 B.C. issued edicts restricting the slaughter of animals and certain kinds of birds, as well as opened veterinary clinics.

Conservation ethics are also found in early religious and philosophical writings. There are examples in the Tao, Shinto, Hindu, Islamic and Buddhist traditions. In Greek philosophy, Plato lamented about pasture land degradation: "What is left now is, so to say, the skeleton of a body wasted by disease; the rich, soft soil has been carried off and only the bare framework of the district left." In the bible, through Moses, God commanded to let the land rest from cultivation every seventh year. Before the 18th century, however, much of European culture considered it a pagan view to admire nature. Wilderness was denigrated while agricultural development was praised. However, as early as AD 680 a wildlife sanctuary was founded on the Farne Islands by St Cuthbert in response to his religious beliefs.

Early Naturalists

Natural history was a major preoccupation in the 18th century, with grand expeditions and the opening of popular public displays in Europe and North America. By 1900 there were 150 natural history museums in Germany, 250 in Great Britain, 250 in the United States, and 300 in France. Preservationist or conservationist sentiments are a development of the late 18th to early 20th centuries.

White gyrfalcons drawn by John James Audubon.

Before Charles Darwin set sail on the HMS Beagle, most people in the world, including Darwin, believed in special creation and that all species were unchanged. George-Louis Leclerc was one of the first naturalist that questioned this belief. He proposed in his 44 volume natural history book that species evolve due to environmental influences. Erasmus Darwin was also a naturalist who also suggested that species evolved. Erasmus Darwin noted that some species have vestigial structures which are anatomical structures that have no apparent function in the species currently but would have been useful for the species' ancestors. The thinking of these early 18th century naturalist helped to change the mindset and thinking of the early 19th century naturalist.

By the early 19th century biogeography was ignited through the efforts of Alexander von Humboldt, Charles Lyell and Charles Darwin. The 19th-century fascination with natural history engendered a fervor to be the first to collect rare specimens with the goal of doing so before they became extinct by other such collectors. Although the work of many 18th and 19th century naturalists were to inspire nature enthusiasts and conservation organizations, their writings, by modern standards, showed insensitivity towards conservation as they would kill hundreds of specimens for their collections.

Conservation Movement

The modern roots of conservation biology can be found in the late 18th-century Enlightenment period particularly in England and Scotland. A number of thinkers, among them notably Lord Monboddo, described the importance of "preserving nature"; much of this early emphasis had its origins in Christian theology.

Scientific conservation principles were first practically applied to the forests of British India. The conservation ethic that began to evolve included three core principles: that human activity damaged the environment, that there was a civic duty to maintain the environment for future generations, and that scientific, empirically based methods should be applied to ensure this duty was carried out. Sir James Ranald Martin was prominent in promoting this ideology, publishing many medico-topographical reports that demonstrated the scale of damage wrought through large-scale deforestation and desiccation, and lobbying extensively for the institutionalization of forest conservation activities in British India through the establishment of Forest Departments.

The Madras Board of Revenue started local conservation efforts in 1842, headed by Alexander Gibson, a professional botanist who systematically adopted a forest conservation program based on scientific principles. This was the first case of state conservation management of forests in the world. Governor-General Lord Dalhousie introduced the first permanent and large-scale forest conservation program in the world in 1855, a model that soon spread to other colonies, as well the United States, where Yellowstone National Park was opened in 1872 as the world's first national park.

The term *conservation* came into widespread use in the late 19th century and referred to the management, mainly for economic reasons, of such natural resources as timber, fish, game, topsoil, pastureland, and minerals. In addition it referred to the preservation of forests (forestry), wildlife (wildlife refuge), parkland, wilderness, and watersheds. This period also saw the passage of the first conservation legislation and the establishment of the first nature conservation societies. The Sea Birds Preservation Act of 1869 was passed in Britain as the first nature protection law in the world after extensive lobbying from the Association for the Protection of Seabirds and the respected ornithologist Alfred Newton. Newton was also instrumental in the passage of the first Game laws from 1872, which protected animals during their breeding season so as to prevent the stock from being brought close to extinction.

One of the first conservation societies was the Royal Society for the Protection of Birds, founded in 1889 in Manchester as a protest group campaigning against the use of great crested grebe and kittiwake skins and feathers in fur clothing. Originally known as "the Plumage League", the group gained popularity and eventually amalgamated with the Fur and Feather League in Croydon, and formed the RSPB. The National Trust formed in 1895 with the manifesto to "...promote the permanent preservation, for the benefit of the nation, of lands, ...to preserve (so far practicable) their natural aspect."

Roosevelt and Muir on Glacier Point in Yosemite National Park.

In the United States, the Forest Reserve Act of 1891 gave the President power to set aside forest reserves from the land in the public domain. John Muir founded the Sierra Club in 1892, and the New York Zoological Society was set up in 1895. A series of national forests and preserves were established by Theodore Roosevelt from 1901 to 1909. The 1916 National Parks Act, included a 'use without impairment' clause, sought by John Muir, which eventually resulted in the removal of a proposal to build a dam in Dinosaur National Monument in 1959.

In the 20th century, Canadian civil servants, including Charles Gordon Hewitt and James Harkin spearheaded the movement toward wildlife conservation.

Global Conservation Efforts

In the mid-20th century, efforts arose to target individual species for conservation, notably efforts in big cat conservation in South America led by the New York Zoological Society. In the early 20th century the New York Zoological Society was instrumental in developing concepts of establishing preserves for particular species and conducting the necessary conservation studies to determine the suitability of locations that are most appropriate as conservation priorities; the work of Henry Fairfield Osborn Jr., Carl E. Akeley, Archie Carr and his son Archie Carr III is notable in this era. Akeley for example, having led expeditions to the Virunga Mountains and observed the mountain gorilla in the wild, became convinced that the species and the area were conservation priorities. He was instrumental in persuading Albert I of Belgium to act in defense of the mountain gorilla and establish Albert National Park (since renamed Virunga National Park) in what is now Democratic Republic of Congo.

By the 1970s, led primarily by work in the United States under the Endangered Species Act along with the Species at Risk Act (SARA) of Canada, Biodiversity Action Plans developed in Australia,

Sweden, the United Kingdom, hundreds of species specific protection plans ensued. Notably the United Nations acted to conserve sites of outstanding cultural or natural importance to the common heritage of mankind. The programme was adopted by the General Conference of UNESCO in 1972. As of 2006, a total of 830 sites are listed: 644 cultural, 162 natural. The first country to pursue aggressive biological conservation through national legislation was the United States, which passed back to back legislation in the Endangered Species Act (1966) and National Environmental Policy Act (1970), which together injected major funding and protection measures to large-scale habitat protection and threatened species research. Other conservation developments, however, have taken hold throughout the world. India, for example, passed the Wildlife Protection Act of 1972.

In 1980, a significant development was the emergence of the urban conservation movement. A local organization was established in Birmingham, UK, a development followed in rapid succession in cities across the UK, then overseas. Although perceived as a grassroots movement, its early development was driven by academic research into urban wildlife. Initially perceived as radical, the movement's view of conservation being inextricably linked with other human activity has now become mainstream in conservation thought. Considerable research effort is now directed at urban conservation biology. The Society for Conservation Biology originated in 1985.

By 1992, most of the countries of the world had become committed to the principles of conservation of biological diversity with the Convention on Biological Diversity; subsequently many countries began programmes of Biodiversity Action Plans to identify and conserve threatened species within their borders, as well as protect associated habitats. The late 1990s saw increasing professionalism in the sector, with the maturing of organisations such as the Institute of Ecology and Environmental Management and the Society for the Environment.

Since 2000, the concept of landscape scale conservation has risen to prominence, with less emphasis being given to single-species or even single-habitat focused actions. Instead an ecosystem approach is advocated by most mainstream conservationists, although concerns have been expressed by those working to protect some high-profile species.

Ecology has clarified the workings of the biosphere; i.e., the complex interrelationships among humans, other species, and the physical environment. The burgeoning human population and associated agriculture, industry, and the ensuing pollution, have demonstrated how easily ecological relationships can be disrupted.

 The last word in ignorance is the man who says of an animal or plant: "What good is it?" If the land mechanism as a whole is good, then every part is good, whether we understand it or not. If the biota, in the course of aeons, has built something we like but do not understand, then who but a fool would discard seemingly useless parts? To keep every cog and wheel is the first precaution of intelligent tinkering.

—*Aldo Leopold, A Sand County Almanac*

Concepts and Foundations

Measuring Extinction Rates

Extinction rates are measured in a variety of ways. Conservation biologists measure and apply statistical measures of fossil records, rates of habitat loss, and a multitude of other variables such as loss of biodiversity as a function of the rate of habitat loss and site occupancy to obtain such estimates. *The Theory of Island Biogeography* is possibly the most significant contribution toward the scientific understanding of both the process and how to measure the rate of species extinction. The current background extinction rate is estimated to be one species every few years.

The measure of ongoing species loss is made more complex by the fact that most of the Earth's species have not been described or evaluated. Estimates vary greatly on how many species actually exist (estimated range: 3,600,000-111,700,000) to how many have received a species binomial (estimated range: 1.5-8 million). Less than 1% of all species that have been described have been studied beyond simply noting its existence. From these figures, the IUCN reports that 23% of vertebrates, 5% of invertebrates and 70% of plants that have been evaluated are designated as endangered or threatened. Better knowledge is being constructed by The Plant List for actual numbers of species.

Systematic Conservation Planning

Systematic conservation planning is an effective way to seek and identify efficient and effective types of reserve design to capture or sustain the highest priority biodiversity values and to work with communities in support of local ecosystems. Margules and Pressey identify six interlinked stages in the systematic planning approach:

1. Compile data on the biodiversity of the planning region

2. Identify conservation goals for the planning region

3. Review existing conservation areas

4. Select additional conservation areas

5. Implement conservation actions

6. Maintain the required values of conservation areas

Conservation biologists regularly prepare detailed conservation plans for grant proposals or to effectively coordinate their plan of action and to identify best management practices (e.g.). Systematic strategies generally employ the services of Geographic Information Systems to assist in the decision making process.

Conservation Physiology: A Mechanistic Approach to Conservation

Conservation physiology was defined by Steven J. Cooke and colleagues as: 'An integrative scientific discipline applying physiological concepts, tools, and knowledge to characterizing biological diversity and its ecological implications; understanding and predicting how organisms, populations, and ecosystems respond to environmental change and stressors; and solving conservation problems across the broad range of taxa (i.e. including microbes, plants, and animals). Physiology is considered in the broadest possible terms to include functional and mechanistic responses at all scales, and conservation includes the development and refinement of strategies to rebuild populations, restore ecosystems, inform conservation policy, generate decision-support tools, and manage natural resources.' Conservation physiology is particularly relevant to practitioners in that it has the potential to generate cause-and-effect relationships and reveal the factors that contribute to population declines.

Conservation Biology as a Profession

The Society for Conservation Biology is a global community of conservation professionals dedicated to advancing the science and practice of conserving biodiversity. Conservation biology as a discipline reaches beyond biology, into subjects such as philosophy, law, economics, humanities, arts, anthropology, and education. Within biology, conservation genetics and evolution are immense fields unto themselves, but these disciplines are of prime importance to the practice and profession of conservation biology.

Is conservation biology an objective science when biologists advocate for an inherent value in nature? Do conservationists introduce bias when they support policies using qualitative description, such as habitat *degradation*, or *healthy* ecosystems? As all scientists hold values, so do conservation biologists. Conservation biologists advocate for reasoned and sensible management of natural resources and do so with a disclosed combination of science, reason, logic, and values in their conservation management plans. This sort of advocacy is similar to the medical profession advocating for healthy lifestyle options, both are beneficial to human well-being yet remain scientific in their approach.

There is a movement in conservation biology suggesting a new form of leadership is needed to mobilize conservation biology into a more effective discipline that is able to communicate the full scope of the problem to society at large. The movement proposes an adaptive leadership approach that parallels an adaptive management approach. The concept is based on a new philosophy or leadership theory steering away from historical notions of power, authority, and dominance. Adaptive conservation leadership is reflective and more equitable as it applies to any member of society who can mobilize others toward meaningful change using communication techniques that are inspiring, purposeful, and collegial. Adaptive conservation leadership and mentoring programs are being implemented by conservation biologists through organizations such as the Aldo Leopold Leadership Program

Approaches

Conservation may be classified as either in-situ conservation, which is protecting an endangered species in its natural habitat, or ex-situ conservation, which occurs outside the natural habitat. In-situ conservation involves protecting or restoring the habitat. Ex-situ conservation, on the other hand, involves protection outside of an organism's natural habitat, such as on reservations or in gene banks, in circumstances where viable populations may not be present in the natural habitat.

Also, non-interference may be used, which is termed a preservationist method. Preservationists advocate for giving areas of nature and species a protected existence that halts interference from the humans. In this regard, conservationists differ from preservationists in the social dimension, as conservation biology engages society and seeks equitable solutions for both society and ecosystems. Some preservationists emphasize the potential of biodiversity in a world without humans.

Ethics and Values

Conservation biologists are interdisciplinary researchers that practice ethics in the biological and social sciences. Chan states that conservationists must advocate for biodiversity and can do so in a scientifically ethical manner by not promoting simultaneous advocacy against other competing values.

A conservationist may be inspired by the *resource conservation ethic*, which seeks to identify what measures will deliver "the greatest good for the greatest number of people for the longest time." In contrast, some conservation biologists argue that nature has an intrinsic value that is independent of anthropocentric usefulness or utilitarianism. Intrinsic value advocates that a gene, or species, be valued because they have a utility for the ecosystems they sustain. Aldo Leopold was a classical thinker and writer on such conservation ethics whose philosophy, ethics and writings are still valued and revisited by modern conservation biologists.

Conservation Priorities

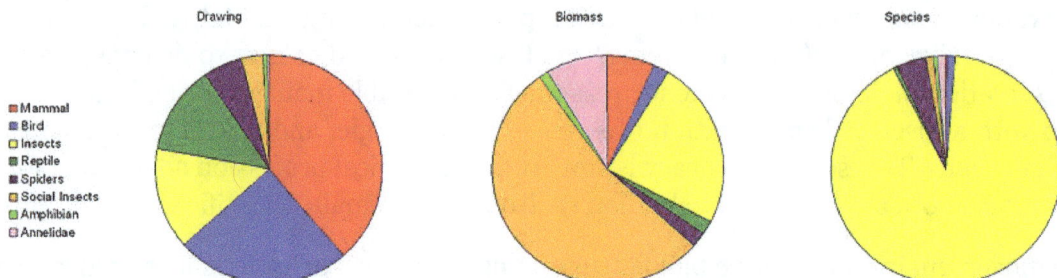

A pie chart image showing the relative biomass representation in a rain forest through a summary of children's perceptions from drawings and artwork (left), through a scientific estimate of actual biomass (middle), and by a measure of biodiversity (right). Notice that the biomass of social insects (middle) far outweighs the number of species (right).

The International Union for the Conservation of Nature (IUCN) has organized a global assortment of scientists and research stations across the planet to monitor the changing state of nature in an effort to tackle the extinction crisis. The IUCN provides annual updates on the status of species conservation through its Red List. The IUCN Red List serves as an international conservation tool to identify those species most in need of conservation attention and by providing a global index on the status of biodiversity. More than the dramatic rates of species loss, however, conservation scientists note that the sixth mass extinction is a biodiversity crisis requiring far more action than a priority focus on rare, endemic or endangered species. Concerns for biodiversity loss covers a broader conservation mandate that looks at ecological processes, such as migration, and a holistic examination of biodiversity at levels beyond the species, including genetic, population and ecosystem diversity. Extensive, systematic, and rapid rates of biodiversity loss threatens the sustained well-being of humanity by limiting supply of ecosystem services that are otherwise regenerated by the complex and evolving holistic network of genetic and ecosystem diversity. While the conservation status of species is employed extensively in conservation management, some scientists highlight that it is the common species that are the primary source of exploitation and habitat alteration by humanity. Moreover, common species are often undervalued despite their role as the primary source of ecosystem services.

While most in the community of conservation science "stress the importance" of sustaining biodiversity, there is debate on how to prioritize genes, species, or ecosystems, which are all components of biodiversity (e.g. Bowen, 1999). While the predominant approach to date has been to focus efforts on endangered species by conserving *biodiversity hotspots*, some scientists (e.g) and conservation organizations, such as the Nature Conservancy, argue that it is more cost-effective, logical, and socially relevant to invest in *biodiversity coldspots*. The costs of discovering, naming, and mapping out the distribution of every species, they argue, is an ill-advised conservation venture. They reason it is better to understand the significance of the ecological roles of species.

Biodiversity hotspots and coldspots are a way of recognizing that the spatial concentration of genes, species, and ecosystems is not uniformly distributed on the Earth's surface. For example, "[…] 44% of all species of vascular plants and 35% of all species in four vertebrate groups are confined to 25 hotspots comprising only 1.4% of the land surface of the Earth."

Those arguing in favor of setting priorities for coldspots point out that there are other measures to consider beyond biodiversity. They point out that emphasizing hotspots downplays the importance of the social and ecological connections to vast areas of the Earth's ecosystems where biomass, not biodiversity, reigns supreme. It is estimated that 36% of the Earth's surface, encompassing 38.9% of the worlds vertebrates, lacks the endemic species to qualify as biodiversity hotspot. Moreover, measures show that maximizing protections for biodiversity does not capture ecosystem services any better than targeting randomly chosen regions. Population level biodiversity (i.e. coldspots) are disappearing at a rate that is ten times that at the species level. The level of importance in addressing biomass versus endemism as a concern for conservation biology is highlighted in literature measuring the level of threat to global ecosystem carbon stocks that do not necessarily reside in areas of endemism. A hotspot priority approach would not invest so heavily in places such as steppes, the Serengeti, the Arctic, or taiga. These areas contribute a great abundance of population (not species) level biodiversity and ecosystem services, including cultural value and planetary nutrient cycling.

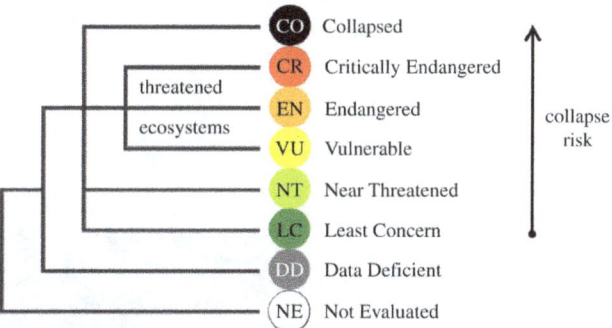

Summary of 2006 IUCN Red List categories.

Those in favor of the hotspot approach point out that species are irreplaceable components of the global ecosystem, they are concentrated in places that are most threatened, and should therefore receive maximal strategic protections. The IUCN Red List categories, species articles, is an example of the hotspot conservation approach in action; species that are not rare or endemic are listed the least concern9. This is a hotspot approach because the priority is set to target species level concerns over population level or biomass. Species richness and genetic biodiversity contributes to and engenders ecosystem stability, ecosystem processes, evolutionary adaptability, and biomass. Both sides agree, however, that conserving biodiversity is necessary to reduce the extinction rate and identify an inherent value in nature; the debate hinges on how to prioritize limited conservation resources in the most cost-effective way.

Economic Values and Natural Capital

Conservation biologists have started to collaborate with leading global economists to determine how to measure the wealth and services of nature and to make these values apparent in global market transactions. This system of accounting is called *natural capital* and would, for example, register the value of an ecosystem before it is cleared to make way for development. The WWF publishes its *Living Planet Report* and provides a global index of biodiversity by monitoring approximately 5,000 populations in 1,686 species of vertebrate (mammals, birds, fish, reptiles, and amphibians) and report on the trends in much the same way that the stock market is tracked.

This method of measuring the global economic benefit of nature has been endorsed by the G8+5 leaders and the European Commission. Nature sustains many ecosystem services that benefit humanity. Many of the earths ecosystem services are public goods without a market and therefore no price or value. When the *stock market* registers a financial crisis, traders on Wall Street are not in the business of trading stocks for much of the planet's living natural capital stored in ecosystems. There is no natural stock market with investment portfolios into sea horses, amphibians, insects, and other creatures that provide a sustainable supply of ecosystem services that are valuable to society. The ecological footprint of society has exceeded the bio-regenerative capacity limits of the planet's ecosystems by about 30 percent, which is the same percentage of vertebrate populations that have registered decline from 1970 through 2005.

The ecological credit crunch is a global challenge. The *Living Planet Report 2008* tells us that more than three-quarters of the world's people live in nations that are ecological debtors – their national consumption has outstripped their country's biocapacity. Thus, most of us are propping up our current lifestyles, and our economic growth, by drawing (and increasingly overdrawing) upon the ecological capital of other parts of the world.

WWF Living Planet Report

Tadrart Acacus desert in western Libya, part of the Sahara

The inherent natural economy plays an essential role in sustaining humanity, including the regulation of global atmospheric chemistry, pollinating crops, pest control, cycling soil nutrients, purifying our water supply, supplying medicines and health benefits, and unquantifiable quality of life improvements. There is a relationship, a correlation, between markets and natural capital, and social income inequity and biodiversity loss. This means that there are greater rates of biodiversity loss in places where the inequity of wealth is greatest.

Although a direct market comparison of natural capital is likely insufficient in terms of human value, one measure of ecosystem services suggests the contribution amounts to trillions of dollars yearly. For example, one segment of North American forests has been assigned an annual value of

250 billion dollars; as another example, honey-bee pollination is estimated to provide between 10 and 18 billion dollars of value yearly. The value of ecosystem services on one New Zealand island has been imputed to be as great as the GDP of that region. This planetary wealth is being lost at an incredible rate as the demands of human society is exceeding the bio-regenerative capacity of the Earth. While biodiversity and ecosystems are resilient, the danger of losing them is that humans cannot recreate many ecosystem functions through technological innovation.

Strategic Species Concepts

Keystone Species

Some species, called a *keystone species* form a central supporting hub unique to their the ecosystem. The loss of such a species results in a collapse in ecosystem function, as well as the loss of coexisting species. Keystone species are usually predators due to their ability to control the population of prey in their ecosystem. The importance of a keystone species was shown by the extinction of the Steller's sea cow (*Hydrodamalis gigas*) through its interaction with sea otters, sea urchins, and kelp. Kelp beds grow and form nurseries in shallow waters to shelter creatures that support the food chain. Sea urchins feed on kelp, while sea otters feed on sea urchins. With the rapid decline of sea otters due to overhunting, sea urchin populations grazed unrestricted on the kelp beds and the ecosystem collapsed. Left unchecked, the urchins destroyed the shallow water kelp communities that supported the Steller's sea cow's diet and hastened their demise. The sea otter was thought to be a keystone species because the coexistence of many ecological associates in the kelp beds relied upon otters for their survival. However this was later questioned by Turvey and Risley, who showed that hunting alone would have driven the Steller's sea cow extinct.

Indicator Species

An *indicator species* has a narrow set of ecological requirements, therefore they become useful targets for observing the health of an ecosystem. Some animals, such as amphibians with their semi-permeable skin and linkages to wetlands, have an acute sensitivity to environmental harm and thus may serve as a *miner's canary*. Indicator species are monitored in an effort to capture environmental degradation through pollution or some other link to proximate human activities. Monitoring an indicator species is a measure to determine if there is a significant environmental impact that can serve to advise or modify practice, such as through different forest silviculture treatments and management scenarios, or to measure the degree of harm that a pesticide may impart on the health of an ecosystem.

Government regulators, consultants, or NGOs regularly monitor indicator species, however, there are limitations coupled with many practical considerations that must be followed for the approach to be effective. It is generally recommended that multiple indicators (genes, populations, species, communities, and landscape) be monitored for effective conservation measurement that prevents harm to the complex, and often unpredictable, response from ecosystem dynamics (Noss, 1997).

Umbrella and Flagship Species

An example of an *umbrella species* is the monarch butterfly, because of its lengthy migrations and aesthetic value. The monarch migrates across North America, covering multiple ecosystems and so requires a large area to exist. Any protections afforded to the monarch butterfly will at the same time umbrella many other species and habitats. An umbrella species is often used as *flagship species*, which are species, such as the giant panda, the blue whale, the tiger, the mountain gorilla and the monarch butterfly, that capture the public's attention and attract support for conservation measures.

Context and Trends

Conservation biologists study trends and process from the paleontological past to the ecological present as they gain an understanding of the context related to species extinction. It is generally accepted that there have been five major global mass extinctions that register in Earth's history. These include: the Ordovician (440 mya), Devonian (370 mya), Permian–Triassic (245 mya), Triassic–Jurassic (200 mya), and Cretaceous–Paleogene extinction event (66 mya) extinction spasms. Within the last 10,000 years, human influence over the Earth's ecosystems has been so extensive that scientists have difficulty estimating the number of species lost; that is to say the rates of deforestation, reef destruction, wetland draining and other human acts are proceeding much faster than human assessment of species. The latest *Living Planet Report* by the World Wide Fund for Nature estimates that we have exceeded the bio-regenerative capacity of the planet, requiring 1.5 Earths to support the demands placed on our natural resources.

Holocene Extinction

An art scape image showing the relative importance of animals in a rain forest through a summary of (a) child's perception compared with (b) a scientific estimate of the importance. The size of the animal represents its importance. The child's mental image places importance on big cats, birds, butterflies, and then reptiles versus the actual dominance of social insects (such as ants).

Conservation biologists are dealing with and have published evidence from all corners of the planet indicating that humanity may be causing the sixth and fastest planetary extinction event. It has been suggested that we are living in an era of unprecedented numbers of species extinctions, also known as the Holocene extinction event. The global extinction rate may be approximately 1,000 times higher than the natural background extinction rate. It is estimated that two-thirds of all mammal genera and one-half of all mammal species weighing at least 44 kilograms (97 lb) have gone extinct in the last 50,000 years. The Global Amphibian Assessment reports that amphibians are declining on a global scale faster than any other vertebrate group, with over 32% of all surviving species being threatened with extinction. The surviving populations are in continual decline in 43% of those that are threatened. Since the mid-1980s the actual rates of extinction have exceeded 211 times rates measured from the fossil record. However, "The current amphibian extinction rate may range from 25,039 to 45,474 times the background extinction rate for amphibians." The global extinction trend occurs in every major vertebrate group that is being monitored. For example, 23% of all mammals and 12% of all birds are Red Listed by the International Union for Conservation of Nature (IUCN), meaning they too are threatened with extinction. Even though extinction is natural, the decline in species is happening at such an incredible rate that evolution can simply not match, therefore, leading to the greatest continual mass extinction on Earth. Humans have dominated the planet and our high consumption of resources, along with the pollution generated is affecting the environments in which other species live. There are a wide variety of species that humans are working to protect such as the Hawaiian Crow and the Whooping Crane of Texas. People can also take action on preserving species by advocating and voting for policies that improve climate. The Earth's oceans especially require attention, as climate change has altered pH levels making it inhabitable for organisms with shells, that are dissolving as a result.

Status of Oceans and Reefs

Global assessments of coral reefs of the world continue to report drastic and rapid rates of decline. By 2000, 27% of the world's coral reef ecosystems had effectively collapsed. The largest period of decline occurred in a dramatic "bleaching" event in 1998, where approximately 16% of all the coral reefs in the world disappeared in less than a year. *Coral bleaching* is caused by a mixture of environmental stresses, including increases in ocean temperatures and acidity, causing both the release of symbiotic algae and death of corals. Decline and extinction risk in coral reef biodiversity has risen dramatically in the past ten years. The loss of coral reefs, which are predicted to go extinct in the next century, threatens the balance of global biodiversity, will have huge economic impacts, and endangers food security for hundreds of millions of people. Conservation biology plays an important role in international agreements covering the world's oceans (and other issues pertaining to biodiversity).

These predictions will undoubtedly appear extreme, but it is difficult to imagine how such changes will not come to pass without fundamental changes in human behavior.

—J.B. Jackson

The oceans are threatened by acidification due to an increase in CO_2 levels. This is a most serious threat to societies relying heavily upon oceanic natural resources. A concern is that the majority of all marine species will not be able to evolve or acclimate in response to the changes in the ocean chemistry.

The prospects of averting mass extinction seems unlikely when "[...] 90% of all of the large (average approximately ≥50 kg), open ocean tuna, billfishes, and sharks in the ocean" are reportedly gone. Given the scientific review of current trends, the ocean is predicted to have few surviving multi-cellular organisms with only microbes left to dominate marine ecosystems.

Groups Other than Vertebrates

Serious concerns also being raised about taxonomic groups that do not receive the same degree of social attention or attract funds as the vertebrates. These include fungal (including lichen-forming species), invertebrate (particularly insect) and plant communities where the vast majority of biodiversity is represented. Conservation of fungi and conservation of insects, in particular, are both of pivotal importance for conservation biology. As mycorrhizal symbionts, and as decomposers and recyclers, fungi are essential for sustainability of forests. The value of insects in the biosphere is enormous because they outnumber all other living groups in measure of species richness. The greatest bulk of biomass on land is found in plants, which is sustained by insect relations. This great ecological value of insects is countered by a society that often reacts negatively toward these aesthetically 'unpleasant' creatures.

One area of concern in the insect world that has caught the public eye is the mysterious case of missing honey bees (*Apis mellifera*). Honey bees provide an indispensable ecological services through their acts of pollination supporting a huge variety of agriculture crops. The use of honey and wax have become vastly used throughout the world. The sudden disappearance of bees leaving empty hives or colony collapse disorder (CCD) is not uncommon. However, in 16-month period from 2006 through 2007, 29% of 577 beekeepers across the United States reported CCD losses in up to 76% of their colonies. This sudden demographic loss in bee numbers is placing a strain on the agricultural sector. The cause behind the massive declines is puzzling scientists. Pests, pesticides, and global warming are all being considered as possible causes.

Another highlight that links conservation biology to insects, forests, and climate change is the mountain pine beetle (*Dendroctonus ponderosae*) epidemic of British Columbia, Canada, which has infested 470,000 km^2 (180,000 sq mi) of forested land since 1999. An action plan has been prepared by the Government of British Columbia to address this problem.

"This impact [*pine beetle epidemic*] converted the forest from a small net carbon sink to a large net carbon source both during and immediately after the outbreak. In the worst year, the impacts resulting from the beetle outbreak in British Columbia were equivalent to 75% of the average annual direct forest fire emissions from all of Canada during 1959–1999."

—*Kurz et al.*

Conservation Biology of Parasites

A large proportion of parasite species are threatened by extinction. A few of them are being eradicated as pests of humans or domestic animals, however, most of them are harmless. Threats include the decline or fragmentation of host populations, or the extinction of host species.

Threats to Biodiversity

Today, many threats to Biodiversity exist. An acronym that can be used to express the top threats of present-day H.I.P.P.O stands for Habitat Loss, Invasive Species, Pollution, Human Population, and Overharvesting. The primary threats to biodiversity are habitat destruction (such as deforestation, agricultural expansion, urban development), and overexploitation (such as wildlife trade). Habitat fragmentation also poses challenges, because the global network of protected areas only covers 11.5% of the Earth's surface. A significant consequence of fragmentation and lack of linked protected areas is the reduction of animal migration on a global scale. Considering that billions of tonnes of biomass are responsible for nutrient cycling across the earth, the reduction of migration is a serious matter for conservation biology.

Human activities are associated directly or indirectly with nearly every aspect of the current extinction spasm.

—Wake and Vredenburg

However, human activities need not necessarily cause irreparable harm to the biosphere. With conservation management and planning for biodiversity at all levels, from genes to ecosystems, there are examples where humans mutually coexist in a sustainable way with nature. Even with the current threats to biodiversity there are ways we can improve the current condition and start anew.

Many of the threats to biodiversity, including disease and climate change, are reaching inside borders of protected areas, leaving them 'not-so protected' (e.g. Yellowstone National Park). Climate change, for example, is often cited as a serious threat in this regard, because there is a feedback loop between species extinction and the release of carbon dioxide into the atmosphere. Ecosystems store and cycle large amounts of carbon which regulates global conditions. In present day, there have been major climate shifts with temperature changes making survival of some species difficult. The effects of global warming add a catastrophic threat toward a mass extinction of global biological diversity. Conservationists have claimed that not all the species can be saved, and they have to decide which their efforts should be used to protect. This concept is known as the Conservation Triage. The extinction threat is estimated to range from 15 to 37 percent of all species by 2050, or 50 percent of all species over the next 50 years. The current extinction rate is 100-100,000 times more rapid today than the last several billion years.

Conservation (Ethic)

Satellite photograph of industrial deforestation in the Tierras Bajas project in eastern Bolivia, using skyline logging and replacement of forests by agriculture.

Conservation is an ethic of resource use, allocation, and protection. Its primary focus is upon maintaining the health of the natural world, its fisheries, habitats, and biological diversity. Secondary focus is on materials conservation, including non-renewable resources such as metals, minerals and fossil fuels, and energy conservation, which is important to protect the natural world. Those who follow the conservation ethic and, especially, those who advocate or work toward conservation goals are termed conservationists.

The terms *conservation* and *preservation* are frequently conflated outside of the academic, scientific, and professional literatures. The US National Park Service offers the following explanation of the important ways in which these two terms represent very different conceptions of environmental protection ethics:

"Conservation and preservation are closely linked and may indeed seem to mean the same thing. Both terms involve a degree of protection, but how that is protection is carried out is the key difference. Conservation is generally associated with the protection of natural resources, while preservation is associated with the protection of buildings, objects, and landscapes. Put simply, *conservation seeks the proper use of nature, while preservation seeks protection of nature from use.*

During the environmental movement of the early 20th century, two opposing factions emerged: conservationists and preservationists. Conservationists sought to regulate human use while preservationists sought to eliminate human impact altogether."

To conserve habitat in terrestrial ecoregions and to stop deforestation is a goal widely shared by many groups with a wide variety of motivations.

To protect sea life from extinction due to overfishing or climate change is another commonly stated goal of conservation — ensuring that "some will be available for future generations" to continue a way of life.

The consumer conservation ethic is sometimes expressed by the *four R's*: " Rethink, Reduce, Recycle, Repair" This social ethic primarily relates to local purchasing, moral purchasing, the sustained, and efficient use of renewable resources, the moderation of destructive use of finite resources, and the prevention of harm to common resources such as air and water quality, the natural functions of a living earth, and cultural values in a built environment.

The principal value underlying most expressions of the conservation ethic is that the natural world has intrinsic and intangible worth along with utilitarian value — a view carried forward by the scientific conservation movement and some of the older Romantic schools of ecology movement.

More Utilitarian schools of conservation seek a proper valuation of local and global impacts of human activity upon nature in their effect upon human well being, now and to posterity. How such values are assessed and exchanged among people determines the social, political, and personal restraints and imperatives by which conservation is practiced. This is a view common in the modern environmental movement.

These movements have diverged but they have deep and common roots in the conservation movement.

In the United States of America, the year 1864 saw the publication of two books which laid the foundation for Romantic and Utilitarian conservation traditions in America. The posthumous publication of Henry David Thoreau's *Walden* established the grandeur of unspoiled nature as a citadel to nourish the spirit of man. From George Perkins Marsh a very different book, *Man and Nature*, later subtitled "The Earth as Modified by Human Action", catalogued his observations of man exhausting and altering the land from which his sustenance derives.

Terminology

The conservation of natural resources is the fundamental problem. Unless we solve that problem, it will avail us little to solve all others.

—*Theodore Roosevelt*

In common usage, the term refers to the activity of systematically protecting natural resources such as forests, including biological diversity. Carl F. Jordan defines the term as:

biological conservation as being a philosophy of managing the environment in a manner that does not despoil, exhaust or extinguish.

While this usage is not new, the idea of biological conservation has been applied to the principles of ecology, biogeography, anthropology, economy and sociology to maintain. biodiversity.

The term "conservation" itself may cover the concepts such as cultural diversity, genetic diversity and the concept of movements environmental conservation, seedbank (preservation of seeds). These are often summarized as the priority to respect diversity, especially by Greens.

Much recent movement in conservation can be considered a resistance to commercialism and globalization. Slow food is a consequence of rejecting these as moral priorities, and embracing a slower and more locally focused lifestyle.

Practice

The Daintree Rainforest in Queensland, Australia

Distinct trends exist regarding conservation development. While many countries' efforts to preserve species and their habitats have been government-led, those in the North Western Europe tended to arise out of the middle-class and aristocratic interest in natural history, expressed at the level of the individual and the national, regional or local learned society. Thus countries like Britain, the Netherlands, Germany, etc. had what we would today term NGOs — in the shape of the RSPB, National Trust and County Naturalists' Trusts (dating back to 1889, 1895 and 1912 respectively) Natuurmonumenten, Provincial conservation Trusts for each Dutch province, Vogelbescherming, etc. — a long time before there were national parks and national nature reserves. This in part reflects the absence of wilderness areas in heavily cultivated Europe, as well as a longstanding interest in laissez-faire government in some countries, like the UK, leaving it as no coincidence that John Muir, the Scottish-born founder of the National Park movement (and hence of government-sponsored conservation) did his sterling work in the USA, where he was the motor force behind the establishment of such NPs as Yosemite and Yellowstone. Nowadays, officially more than 10 percent of the world is legally protected in some way or the other, and in practice private fundraising is insufficient to pay for the effective management of so much land with protective status.

Protected areas in developing countries, where probably as many as 70–80 percent of the species of the world live, still enjoy very little effective management and protection. Some countries, such as Mexico, have non-profit civil organizations and land owners dedicated to protect vast private property, such is the case of Hacienda Chichen's Maya Jungle Reserve and Bird Refuge in Chichen Itza, Yucatán. The Adopt A Ranger Foundation has calculated that worldwide about 140,000 rangers are needed for the protected areas in developing and transition countries. There are no data on how many rangers are employed at the moment, but probably less than half the protected areas in developing and transition countries have any rangers at all and those that have them are at least 50% short This means that there would be a worldwide ranger deficit of 105,000 rangers in the developing and transition countries.

One of the world's foremost conservationists, Dr. Kenton Miller, stated about the importance of rangers: "The future of our ecosystem services and our heritage depends upon park rangers. With the rapidity at which the challenges to protected areas are both changing and increasing, there has never been more of a need for well prepared human capacity to manage. Park rangers are the backbone of park management. They are on the ground. They work on the front line with scientists, visitors, and members of local communities."

Adopt A Ranger, fears that the ranger deficit is the greatest single limiting factor in effectively conserving nature in 75% of the world. Currently, no conservation organization or western country or international organization addresses this problem. Adopt A Ranger has been incorporated to draw worldwide public attention to the most urgent problem that conservation is facing in developing and transition countries: protected areas without field staff. Very specifically, it will contribute to solving the problem by fund raising to finance rangers in the field. It will also help governments in developing and transition countries to assess realistic staffing needs and staffing strategies.

Others, including Survival International, have advocated instead for cooperation with local tribal peoples, who are natural allies of the conservation movement and can provide cost-effective protection.

Habitat Conservation

Habitat conservation is a management practice that seeks to conserve, protect and restore habitat areas for wild plants and animals, especially conservation reliant species, and prevent their extinction, fragmentation or reduction in range. It is a priority of many groups that cannot be easily characterized in terms of any one ideology.

Tree planting is an aspect of habitat conservation. In each plastic tube a hardwood tree has been planted.

There are significant ecological benefits associated with selective cutting. Pictured is an area with Ponderosa Pine trees that were selectively harvested.

History of the Conservation Movement

For much of human history, *nature* had been seen as a resource, one that could be controlled by the government and used for personal and economic gain. The idea was that plants only existed to feed animals and animals only existed to feed humans. The land itself had limited value only extending to the resources it could provide such as minerals and oil.

Throughout the 18th and 19th centuries social views started to change and scientific conservation principles were first practically applied to the forests of British India. The conservation ethic that began to evolve included three core principles: that human activity damaged the environment, that there was a civic duty to maintain the environment for future generations, and that scientific, empirically based methods should be applied to ensure this duty was carried out. Sir James Ranald Martin was prominent in promoting Pbnjbthis ideology, publishing many medico-topographical reports that demonstrated the scale of damage wrought through large-scale deforestation and desiccation, and lobbying extensively for the institutionalization of forest conservation activities in British India through the establishment of Forest Departments.

The Madras Board of Revenue started local conservation efforts in 1842, headed by Alexander Gibson, a professional botanist who systematically adopted a forest conservation program based on scientific principles. This was the first case of state conservation management of forests in the world. Governor-General Lord Dalhousie introduced the first permanent and large-scale forest conservation program in the world in 1855, a model that soon spread to other colonies, as well the United States, where Yellowstone National Park was opened in 1872 as the world's first national park.

Rather than focusing on the economic or material benefits associated with nature, humans began to appreciate the value of nature itself and the need to protect pristine wilderness. By the middle of the 20th century countries such as the United States, Canada, and Britain understood this appreciation and instigated laws and legislation in order to ensure that the most fragile and beautiful environments would be protected for generations to come. Today with the help of NGO's, not-for profit organizations and governments worldwide there is a stronger movement taking place, with a deeper understanding of habitat conservation with the aim of protecting delicate habitats and preserving biodiversity on a global scale. The commitment and actions of small volunteering association in villages and towns, that endeavour to emulate the work done by well known Conservation Organisations, is paramount in ensuring generations that follow understand the importance of conserving natural resources. A village conservation group with the mission statement "We are committed to protecting and enhancing the natural environment in and around the adjoining villages of Ouston and Urpeth." may one day inspire a child who becomes the employee of a worldwide conservation organisation.

Values of Natural Habitat

The natural environment is a source for a wide range of resources that can be exploited for economic profit, for example timber is harvested from forests and clean water is obtained from natural streams. However, land development from anthropogenic economic growth often causes a decline in the ecological integrity of nearby natural habitat. For instance, this was an issue in the northern rocky mountains of the USA.

However, there is also economic value in conserving natural habitats. Financial profit can be made from tourist revenue, for example in the tropics where species diversity is high, or in recreational sports which take place in natural environments such as hiking and mountain biking. The cost of repairing damaged ecosystems is considered to be much higher than the cost of conserving natural ecosystems.

Measuring the worth of conserving different habitat areas is often criticized as being too utilitarian from a philosophical point of view.

Biodiversity

Habitat conservation is important in maintaining biodiversity, an essential part of global food security. There is evidence to support a trend of accelerating erosion of the genetic resources of agricultural plants and animals. An increase in genetic similarity of agricultural plants and animals means an increased risk of food loss from major epidemics. Wild species of agricultural plants have been found to be more resistant to disease, for example the wild corn species Teosinte is resistant to 4 corn diseases that affect human grown crops. A combination of seed banking and habitat conservation has been proposed to maintain plant diversity for food security purposes.

Classifying Environmental Values

Pearce and Moran outlined the following method for classifying environmental uses:

- Direct extractive uses: e.g. timber from forests, food from plants and animals

- Indirect uses: e.g. ecosystem services like flood control, pest control, erosion protection

- Optional uses: future possibilities e.g. unknown but potential use of plants in chemistry/ medicine

- Non-use values:

  o Bequest value (benefit of an individual who knows that others may benefit from it in future)

  o Passive use value (sympathy for natural environment, enjoyment of the mere existence of a particular species)

Impacts

Natural Causes

Habitat loss and destruction can occur both naturally and through anthropogenic causes. Events leading to natural habitat loss include climate change, catastrophic events such as volcanic explosions and through the interactions of invasive and non-invasive species. Natural climate change, events have previously been the cause of many widespread and large scale losses in habitat. For example, some of the mass extinction events generally referred to as the "Big Five" have coincided with large scale such as the Earth entering an ice age, or alternate warming events. Other events in the big five also have their roots in natural causes, such as volcanic explosions and meteor collisions. The Chicxulub impact is one such example, which has previously caused widespread losses in habitat as the Earth either received less sunlight or grew colder, causing certain fauna and flora to flourish whilst others perished. Previously known warm areas in the tropics, the most sensitive habitats on Earth, grew colder, and areas such as Australia developed radically different flora and fauna to those seen today. The big five mass extinction events have also been linked to sea level changes, indicating that large scale marine species loss was strongly influenced by loss in marine habitats, particularly shelf habitats. Methane-driven oceanic eruptions have also been shown to have caused smaller mass extinction events.

Human Impacts

Humans have been the cause of many species' extinction. Due to humans' changing and modifying their environment, the habitat of other species often become altered or destroyed as a result of human actions. Even before the modern industrial era, humans were having widespread, and major effects on the environment. A good example of this is found in Aboriginal Australians and Australian megafauna. Aboriginal hunting practices, which included burning large sections of forest at a time, eventually altered and changed Australia's vegetation so much that many herbivorous megafauna species were left with no habitat and were driven into extinction. Once herbivorous megafauna species became extinct, carnivorous megafauna species soon followed. In the recent past, humans have been responsible for causing more extinctions within a given period of time than ever before. Deforestation, pollution, anthropogenic climate change and human settlements have all been driving forces in altering or destroying habitats. The destruction of ecosystems such as rainforests has resulted in countless habitats being destroyed. These biodiversity hotspots are home to millions of habitat specialists, which do not exist beyond a tiny area. Once their habitat is destroyed, they cease to exist.This destruction has a follow-on effect, as species which coexist or depend upon the existence of other species also become extinct, eventually resulting in the collapse of an entire ecosystem. These time-delayed extinctions are referred to as the extinction debt, which is the result of destroying and fragmenting habitats. As a result of anthropogenic modification of the environment, the extinction rate has climbed to the point where the Earth is now within a sixth mass extinction event, as commonly agreed by biologists. This has been particularly evident, for example, in the rapid decline in the number of amphibian species worldwide.

Approaches and Methods of Habitat Conservation

Determining the size, type and location of habitat to conserve is a complex area of conservation biology. Although difficult to measure and predict, the conservation value of a habitat is often a reflection of the quality (e.g. species abundance and diversity), endangerment of encompassing ecosystems, and spatial distribution of that habitat.

Identifying Priority Habitats for Conservation

Habitat conservation is vital for protecting species and ecological processes. It is important to conserve and protect the space/ area in which that species occupies. Therefore, areas classified as 'biodiversity hotspots', or those in which a flagship, umbrella, or endangered species inhabits are often the habitats that are given precedence over others. Species that possess an elevated risk of extinction are given the highest priority and as a result of conserving their habitat, other species in that community are protected thus serving as an element of gap analysis. In the United States of America, a Habitat Conservation Plan (HCP) is often developed to conserve the environment in which a specific species inhabits. Under the U.S. Endangered Species Act (ESA) the habitat that requires protection in an HCP is referred to as the 'critical habitat'. Multiple-species HCPs are becoming more favourable than single-species HCPs as they can potentially protect an array of species before they warrant listing under the ESA, as well as being able to conserve broad ecosystem components and processes . As of January 2007, 484 HCPs were permitted across the United States, 40 of which covered 10 or more species.The San Diego Multiple Species Conservation Plan (MSCP) encompasses 85 species in a total area of 26,000-km2. Its aim is to protect the habitats of multiple species and overall biodiversity by minimizing development in sensitive areas.

HCPs require clearly defined goals and objectives, efficient monitoring programs, as well as successful communication and collaboration with stakeholders and land owners in the area. Reserve design is also important and requires a high level of planning and management in order to achieve the goals of the HCP. Successful reserve design often takes the form of a hierarchical system with the most valued habitats requiring high protection being surrounded by buffer habitats that have a lower protection status. Like HCPs, hierarchical reserve design is a method most often used to protect a single species, and as a result habitat corridors are maintained, edge effects are reduced and a broader suite of species are protected.

How Much Habitat is Needed

A range of methods and models currently exist that can be used to determine how much habitat is to be conserved in order to sustain a viable population, including Resource Selection Function and Step Selection models. Modelling tools often rely on the spatial scale of the area as an indicator of conservation value. There has been an increase in emphasis on conserving few large areas of habitat as opposed to many small areas. This idea is often referred to as the "single large or several small", SLOSS debate, and is a highly controversial area among conservation biologists and ecologists. The reasons behind the argument that "larger is better" include the reduction in the negative impacts of patch edge effects, the general idea that species richness increases with habitat area and the ability of larger habitats to support greater populations with lower extinction probabilities. Noss & Cooperrider support the "larger is better" claim and developed a model that implies areas of habitat less than 1000ha are "tiny" and of low conservation value. However, Shwartz suggests that although "larger is better", this does not imply that "small is bad". Shwartz argues that human induced habitat loss leaves no alternative to conserving small areas. Furthermore, he suggests many endangered species which are of high conservation value, may only be restricted to small isolated patches of habitat, and thus would be overlooked if larger areas were given a higher priority. The shift to conserving larger areas is somewhat justified in society by placing more value on larger vertebrate species, which naturally have larger habitat requirements.

Examples of Current Conservation Organizations

The Nature Conservancy

Since its formation in 1951 The Nature Conservancy has slowly developed into one of the world's largest conservation organizations. Currently operating in over 30 countries, across 5 continents worldwide, The Nature Conservancy aims to protect nature and its assets for future generations. The organization purchases land or accepts land donations with the intension of conserving its natural resources. In 1955 The Nature Conservancy purchased its first 60-acre plot near the New York/Connecticut border in the United States of America. Today the Conservancy has expanded to protect over 119 million acres of land, 5,000 river miles as well as participating in over 1000 marine protection programs across the globe . Since its beginnings The Nature Conservancy has understood the benefit in taking a scientific approach towards habitat conservation. For the last decade the organization has been using a collaborative, scientific method known as 'Conservation by Design'. By collecting and analyzing scientific data The Conservancy is able to holistically approach the protection of various ecosystems. This process determines the habitats that need protection, specific elements that should be conserved as well as monitoring progress so more efficient practices can be developed for the future.

The Nature Conservancy currently has a large number of diverse projects in operation. They work with countries around the world to protect forests, river systems, oceans, deserts and grasslands. In all cases the aim is to provide a sustainable environment for both the plant and animal life forms that depend on them as well as all future generations to come. turtles

World Wildlife Fund (WWF)

The World Wildlife Fund (WWF) was first formed in after a group of passionate conservationists signed what is now referred to as the Morges Manifesto. WWF is currently operating in over 100 countries across 5 continents with a current listing of over 5 million supporters. One of the first projects of WWF was assisting in the creation of the Charles Darwin Research Foundation which aided in the protection of diverse range of unique species existing on the Galápagos' Islands, Ecuador. It was also a WWF grant that helped with the formation of the College of African Wildlife Management in Tanzania which today focuses on teaching a wide range of protected area management skills in areas such as ecology, range management and law enforcement. The WWF has since gone on to aid in the protection of land in Spain, creating the Coto Doñana National Park in order to conserve migratory birds and The Democratic Republic of Congo, home to the world's largest protected wetlands. The WWF also initiated a debt-for-nature concept which allows the country to put funds normally allocated to paying off national debt, into conservation programs that protect its natural landscapes. Countries currently participating include Madagascar, the first country to participate which since 1989 has generated over $US50 million towards preservation, Bolivia, Costa Rica, Ecuador, Gabon, the Philippines and Zambia.

Rare Conservation

Rare has been in operation since 1973 with current global partners in over 50 countries and offices in the United States of America, Mexico, the Philippines, China and Indonesia. Rare focuses on the human activity that threatens biodiversity and habitats such as overfishing and unsustainable agriculture. By engaging local communities and changing behaviour Rare has been able to launch campaigns to protect areas in most need of conservation. The key aspect of Rare's methodology is their "Pride Campaign's". For example, in the Andes in South America, Rare has incentives to develop watershed protection practices. In the Southeast Asia's "coral triangle" Rare is training fishers in local communities to better manage the areas around the coral reefs in order to lessen human impact. Such programs last for three years with the aim of changing community attitudes so as to conserve fragile habitats and provide ecological protection for years to come.

WWF Netherlands

WWF Netherlands, along with ARK Nature, Wild Wonders of Europe and Conservation Capital have started the Rewilding Europe project. This project intents to rewild several areas in Europe.

Nest Box

A nest box, also spelled nestbox, is a man-made enclosure provided for animals to nest in. Nest boxes are most frequently utilized for birds, in which case they are also called birdhouses or a birdbox/bird box, but some mammalian species may also use them. Placing nestboxes or roosting boxes may also be used to help maintain populations of particular species in an area. The nest box was invented by the British conservationist Charles Waterton in the early 19th century to encourage more birdlife and wildfowl on the nature reserve he set up on his estate.

Western bluebird leaving a nest box

Construction

General Construction

Nest boxes are usually wooden, although the purple martin will nest in metal. Some boxes are made from a mixture of wood and concrete, called *woodcrete*.

Nest boxes should be made from untreated wood with an overhanging, sloped roof, a recessed floor, drainage and ventilation holes, a way to access the interior for monitoring and cleaning, and have no outside perches which could assist predators. Boxes may either have an entrance hole or be open-fronted. Some nest boxes can be highly decorated and complex, sometimes mimicking human houses or other structures. They may also contain nest box cameras so that use of, and activity within, the box can be monitored.

Bird Nest Box Construction

Birdhouses in Gramercy Park, New York City, note the use of different diameter entrance holes.

The diameter of the opening in a nest-box has a very strong influence on the species of birds that will use the box. Many small birds select boxes with a hole only just large enough for an adult bird to pass through. This may be an adaptation to prevent other birds from raiding it. In European countries, an opening of 2.5 cm in diameter will attract *Poecile palustris*, *Poecile montanus*; an opening of 2.8 cm in diameter will attract *Ficedula hypoleuca*, and an opening of 3 cm in diameter will attract *Parus major*, *Passer montanus*, an opening of 3 cm in diameter will attract *Passer domesticus*.

The size of the nest box also affects the bird species likely to use the box. Very small boxes attract wrens and treecreepers and very large ones may attract ducks and owls. Seasonally removing old nest material and parasites is important if they are to be successfully re-used.

The material used in the construction may also be significant. Sparrows have been shown to prefer woodcrete boxes rather than wooden ones. Birds nesting in woodcrete sites had earlier clutches, a shorter incubation period, and more reproductive success, perhaps because the synthetic nests were warmer than their wooden counterparts.

Placement of the nest box obviously is also significant. Some birds (including birds of prey) prefer their nest box to be at an optimum height. Some birds (such as ducks) prefer nest sites them to be very low or even at ground level. For many birds orientation relative to the sun is of importance with many birds preferring an orientation away from direct sun and sheltered from the prevailing rain.

Bat Box Construction

A typical bat house affixed to a tree trunk.

Bat boxes differ from bird nest-boxes in typical design, with the larger opening on the underside of the box, and are more often referred to as bat boxes, although in regard to the rearing of young, they serve the same purpose. Some threatened bat species can be locally supported with the provision of appropriately placed bat-boxes, however species that roost in foliage or large cavities will not use bat boxes. Bat boxes are typically made out of wood, and there are several designs for boxes with single or multiple chambers. Directions for making the open bottom bat houses for small and large colonies, as well as locations to purchase them are available on the internet. Colour and placement is important to ensuring that bat boxes are used; bat boxes that are too shaded will not heat up enough to attract a maternity colony of bats. Australian bat box projects have been running for over 12 years in particular at the Organ Pipes National Park. Currently there are 42 roost boxes using the "Stebbings Design" which have peaked at 280 bats roosting in them. The biggest problem with roosting boxes of any kind is the ongoing maintenance; problems include boxes falling down, wood deteriorating, and pests such as ants, the occasional rat, possums, and spiders.

Other Creatures

Two wasp nests inside a nest box set for boreal owls.

Nest boxes are marketed not only for birds, but also for butterflies and other mammals, especially arboreal ones such as squirrels and opossums. Depending on the animal, these boxes are used for roosting, breeding, or both. Or, as in the case with butterflies, hibernation.

Wasps may build their nests inside a nest box intended for other animals, and may exclude the intended species.

Wildlife Garden

Joe-Pye weed in flower

A wildlife garden (or wild garden) is an environment created by a gardener that serves as a sustainable haven for surrounding wildlife. Wildlife gardens contain a variety of habitats that cater to native and local plants, birds, amphibians, reptiles, insects, mammals and so on. Establishing a garden environment that mimics surrounding wildlife allows for natural systems to interact and establish an equilibrium, ultimately minimizing the need for gardener maintenance and intervention. Wildlife gardens can also play an essential role in biological pest control, and also promote biodiversity, native plantings, and generally benefit the wider environment.

Habitats

Building a successful garden suitable for local wildlife is best accomplished through the use of multiple three-dimensional habitats with diverse structures that provide places for animals to nest and hide. Wildlife gardens may contain a range of habitats, including:

Log piles – Preferably located in a shady area, a pile of logs is a sanctuary for insects and amphibians. The organic structure is a shelter for both protection and breeding. In addition to logs, garden debris may also be added around the garden to be used as a natural mulch, fertilizer, weed control, soil amendment, and habitat for arthropod predators.

Bird feeding stations and bird houses – A place for birds to eat and take shelter will increase the number of birds in the garden, which play a key role in biological pest control. Not only will food and shelter increase the survival rate of birds, but it will also ensure that they are healthy enough for a successful breeding season.

Bug boxes – Offcuts of wood placed in a structure above ground provides an alternate place of shelter for beneficial insects, such as the robber fly, which help keep natural ecosystem predators in check.

Sources of water – A water feature, such as a pond, has the potential to support a large biodiversity of wildlife. To maximize the amount of wildlife attracted to the water feature, it should consist of ranging depths. Shallow areas are used by birds to drink and by insects and amphibians to lay eggs. Deeper areas provide habitat for aquatic insects and a place for amphibians, or even fish to swim.

Pollinating flowers – Flowers rich in nectar will attract bees and butterflies into the garden. Wildflower meadows are an alternative option for lawns in the garden and will serve as a sanctuary for pollinators. However, pollinating plants should not be confused with plants suitable for butterfly breeding.

Plant diversity – The garden should include a range of plant types to serve different species habitats. A balance between ground cover, shrub, understory, and canopy species will allow different sized wildlife shelters that fit their individual needs.

Choice of plants

Although some exotics may also be included, wild gardens usually mostly feature a variety of native species. Generally, these will be a part of the pre-existing natural ecology of an area, making them easier to grow than most exotic species. Choosing native plants comes with an array of benefits for both plant and animal diversity, especially the ability to support native insect and mushroom populations that have established balanced evolutionary relationships over thousands of years.

Ornamental plants on the market tend to lean toward "pest-free" plants, making it hard for native insects to adapt, and ultimately reducing their food supply. Decreases in insect populations due to excessive ornamental planting will discourage bird populations from inhabiting the area.

Invasive species can always prove problematic in the garden due to the absence of natural predators and their ability to reproduce rapidly. Without any measures of control, invasive species can easily overtake native species in the garden. Addressing invasive plants can be done a variety of ways; however, to ensure the least amount of damage to the surrounding ecosystem, this is best done by cutting down the plant, followed by painting its stem with an herbicide, such as Roundup. The debris from the invasive species can be piled and used as a home for smaller critters.

Wildlife Gardens in the Netherlands

Wildlife gardens in the Netherlands are called "heemtuinen". The first was created in 1925: Thijsse's Hof (Garden of Thijsse) in Bloemendaal, near Haarlem. It was gifted to Jac. P. Thijsse on the occasion of his 60th anniversary, and still exists today. The garden gives a display of about 800 plants native to the dune region of South Kennemerland, in which the garden is situated. It is said to be one of the oldest wildlife gardens of its sort in the world.

Nowadays some 25 wildlife gardens exist in the Netherlands.

Conservation Movement

The conservation movement, also known as nature conservation, is a political, environmental and a social movement that seeks to protect natural resources including animal and plant species as well as their habitat for the future.

The early conservation movement included fisheries and wildlife management, water, soil conservation and sustainable forestry. The contemporary conservation movement has broadened from the early movement's emphasis on use of sustainable yield of natural resources and preservation of wilderness areas to include preservation of biodiversity. Some say the conservation movement is part of the broader and more far-reaching environmental movement, while others argue that they differ both in ideology and practice. Chiefly in the United States, conservation is seen as differing from environmentalism in that it aims to preserve natural resources expressly for their continued sustainable use by humans. In other parts of the world conservation is used more broadly to include the setting aside of natural areas and the active protection of wildlife for their inherent value, as much as for any value they may have for humans.

History

Early History

The conservation movement can be traced back to John Evelyn's work *Sylva*, presented as a paper to the Royal Society in 1662. Published as a book two years later, it was one of the most highly influential texts on forestry ever published. Timber resources in England were becoming dangerously depleted at the time, and Evelyn advocated the importance of conserving the forests by managing the rate of depletion and ensuring that the cut down trees get replenished.

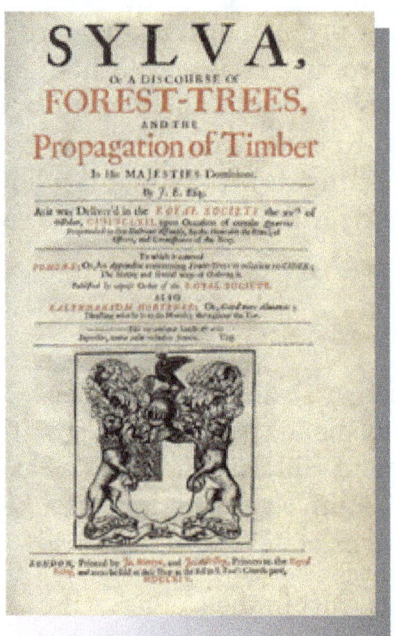

Sylva, or A Discourse of Forest-Trees and the Propagation of Timber in His Majesty's Dominions.

The field developed during the 18th century, especially in Prussia and France where scientific forestry methods were developed. These methods were first applied rigorously in British India from the early-19th century. The government was interested in the use of forest produce and began managing the forests with measures to reduce the risk of wildfire in order to protect the "household" of nature, as it was then termed. This early ecological idea was in order to preserve the growth of delicate teak trees, which was an important resource for the Royal Navy. Concerns over teak depletion were raised as early as 1799 and 1805 when the Navy was undergoing a massive expansion during the Napoleonic Wars; this pressure led to the first formal conservation Act, which prohibited the felling of small teak trees. The first forestry officer was appointed in 1806 to regulate and preserve the trees necessary for shipbuilding. This promising start received a setback in the 1820s and 30s, when laissez-faire economics and complaints from private landowners brought these early conservation attempts to an end.

Origins of the Modern Conservation Movement

Conservation was revived in the mid-19th century, with the first practical application of scientific conservation principles to the forests of India. The conservation ethic that began to evolve included three core principles: that human activity damaged the environment, that there was a civic duty to maintain the environment for future generations, and that scientific, empirically based methods should be applied to ensure this duty was carried out. Sir James Ranald Martin was prominent in promoting this ideology, publishing many medico-topographical reports that demonstrated the scale of damage wrought through large-scale deforestation and desiccation, and lobbying extensively for the institutionalization of forest conservation activities in British India through the establishment of Forest Departments. Edward Percy Stebbing warned of desertification of India. The Madras Board of Revenue started local conservation efforts in 1842, headed by Alexander Gibson, a professional botanist who systematically adopted a forest conservation program based on scientific principles. This was the first case of state management of forests in the world.

These local attempts gradually received more attention by the British government as the unregulated felling of trees continued unabated. In 1850, the British Association in Edinburgh formed a committee to study forest destruction at the behest of Dr. Hugh Cleghorn a pioneer in the nascent conservation movement.

He had become interested in forest conservation in Mysore in 1847 and gave several lectures at the Association on the failure of agriculture in India. These lectures influenced the government under Governor-General Lord Dalhousie to introduce the first permanent and large-scale forest conservation program in the world in 1855, a model that soon spread to other colonies, as well the United States. In the same year, Cleghorn organised the Madras Forest Department and in 1860 the Department banned the use shifting cultivation. Cleghorn's 1861 manual, *The forests and gardens of South India*, became the definitive work on the subject and was widely used by forest assistants in the subcontinent. In 1861, the Forest Department extended its remit into the Punjab.

Schlich, in the middle of the seated row, with students from the forestry school
at Oxford, on a visit to the forests of Saxony in the year 1892.

Sir Dietrich Brandis, a German forester, joined the British service in 1856 as superintendent of the teak forests of Pegu division in eastern Burma. During that time Burma's teak forests were controlled by militant Karen tribals. He introduced the "taungya" system, in which Karen villagers provided labour for clearing, planting and weeding teak plantations. After seven years in Burma, Brandis was appointed Inspector General of Forests in India, a position he served in for 20 years. He formulated new forest legislation and helped establish research and training institutions. The Imperial Forest School at Dehradun was founded by him.

Germans were prominent in the forestry administration of British India. As well as Brandis, Berthold Ribbentrop and Sir William P.D. Schlich brought new methods to Indian conservation, the latter becoming the Inspector-General in 1883 after Brandis stepped down. Schlich helped to establish the journal *Indian Forester* in 1874, and became the founding director of the first forestry school in England at Cooper's Hill in 1885. He authored the five-volume *Manual of Forestry*

(1889–96) on silviculture, forest management, forest protection, and forest utilisation, which became the standard and enduring textbook for forestry students.

Conservation in the United States

F. V. Hayden's map of Yellowstone National Park, 1871.

The American movement received its inspiration from 19th century works that exalted the inherent value of nature, quite apart from human usage. Author Henry David Thoreau (1817-1862) made key philosophical contributions that exalted nature. Thoreau was interested in peoples' relationship with nature and studied this by living close to nature in a simple life. He published his experiences in the book *Walden,* which argued that people should become intimately close with nature. The ideas of Sir Brandis, Sir William P.D. Schlich and Carl A. Schenck were also very influential - Gifford Pinchot, the first chief of the USDA Forest Service, relied heavily upon Brandis' advice for introducing professional forest management in the U.S. and on how to structure the Forest Service.

Both Conservationists and Preservationists appeared in political debates during the Progressive Era (the 1890s—early 1920s). There were three main positions. The laissez-faire position held that owners of private property—including lumber and mining companies, should be allowed to do anything they wished for their property.

The conservationists, led by future President Theodore Roosevelt and his close ally George Bird Grinnell, were motivated by the wanton waste that was taking place at the hand of market forces, including logging and hunting. This practice resulted in placing a large number of North American game species on the edge of extinction. Roosevelt recognized that the laissez-faire approach of the U.S. Government was too wasteful and inefficient. In any case, they noted, most of the natural resources in the western states were already owned by the federal government. The best course of action, they argued, was a long-term plan devised by national experts to maximize the long-term economic benefits of natural resources. To accomplish the mission, Roosevelt and Grinnell formed the Boone and Crockett Club in 1887. The Club was made up of the best minds and influential men of the day. The Boone and Crockett Club's contingency of conservationists, scientists, politicians, and intellectuals became Roosevelt's closest advisers during his march to preserve wildlife and habitat across North America. Preservationists, led by John Muir (1838–1914), argued that the conservation policies were not strong enough to protect the interest of the natural world because they continued to focus on the natural world as a source of economic production.

The debate between conservation and preservation reached its peak in the public debates over the construction of California's Hetch Hetchy dam in Yosemite National Park which supplies the water supply of San Francisco. Muir, leading the Sierra Club, declared that the valley must be preserved for the sake of its beauty: "No holier temple has ever been consecrated by the heart of man."

President Roosevelt put conservationist issue high on the national agenda. He worked with all the major figures of the movement, especially his chief advisor on the matter, Gifford Pinchot and was deeply committed to conserving natural resources. He encouraged the Newlands Reclamation Act of 1902 to promote federal construction of dams to irrigate small farms and placed 230 million acres (360,000 mi² or 930,000 km²) under federal protection. Roosevelt set aside more federal land for national parks and nature preserves than all of his predecessors combined.

Gifford Pinchot had been appointed by McKinley as chief of Division of Forestry in the Department of Agriculture. In 1905, his department gained control of the national forest reserves. Pinchot promoted private use (for a fee) under federal supervision. In 1907, Roosevelt designated 16 million acres (65,000 km²) of new national forests just minutes before a deadline.

A PRACTICAL FORESTER
(A subject that had attention all through Mr.

Roosevelt established the United States Forest Service, signed into law the creation of five national parks, and signed the year 1906 Antiquities Act, under which he proclaimed 18 new national monuments. He also established the first 51 bird reserves, four game preserves, and 150 national forests, including Shoshone National Forest, the nation's first. The area of the United States that he placed under public protection totals approximately 230,000,000 acres (930,000 km²).

In May 1908, Roosevelt sponsored the Conference of Governors held in the White House, with a focus on natural resources and their most efficient use. Roosevelt delivered the opening address: "Conservation as a National Duty".

In 1903 Roosevelt toured the Yosemite Valley with John Muir, who had a very different view of conservation, and tried to minimize commercial use of water resources and forests. Working through the Sierra Club he founded, Muir succeeded in 1905 in having Congress transfer the Mariposa Grove and Yosemite Valley to the federal government. While Muir wanted nature preserved for its own sake, Roosevelt subscribed to Pinchot's formulation, "to make the forest produce the largest amount of whatever crop or service will be most useful, and keep on producing it for generation after generation of men and trees."

Theodore Roosevelt's view on conservationism remained dominant for decades; - Franklin D. Roosevelt authorised the building of many large-scale dams and water projects, as well as the expansion of the National Forest System to buy out sub-marginal farms. In 1937, the Pittman–Robertson Federal Aid in Wildlife Restoration Act was signed into law, providing funding for state agencies to carry out their conservation efforts.

Since 1970

Environmental reemerged on the national agenda in 1970, with Republican Richard Nixon playing a major role, especially with his creation of the Environmental Protection Agency. The debates over the public lands and environmental politics played a supporting role in the decline of liberalism and the rise of modern environmentalism. Although Americans consistently rank environmental issues as "important", polling data indicates that in the voting booth voters rank the environmental issues low relative to other political concerns.

The growth of the Republican party's political power in the inland West (apart from the Pacific coast) was facilitated by the rise of popular opposition to public lands reform. Successful Democrats in the inland West and Alaska typically take more conservative positions on environmental issues than Democrats from the Coastal states. Conservatives drew on new organizational networks of think tanks, industry groups, and citizen-oriented organizations, and they began to deploy new strategies that affirmed the rights of individuals to their property, protection of extraction rights, to hunt and recreate, and to pursue happiness unencumbered by the federal government at the expense of resource conservation.

Areas of Concern

Deforestation and overpopulation are issues affecting all regions of the world. The consequent destruction of wildlife habitat has prompted the creation of conservation groups in other countries, some founded by local hunters who have witnessed declining wildlife populations first hand. Also, it was highly important for the conservation movement to solve problems of living conditions in the cities and the overpopulation of such places.

Boreal Forest and the Arctic

The idea of incentive conservation is a modern one but its practice has clearly defended some of the sub Arctic wildernesses and the wildlife in those regions for thousands of years, especially by indigenous peoples such as the Evenk, Yakut, Sami, Inuit and Cree. The fur trade and hunting by these peoples have preserved these regions for thousands of years. Ironically, the pressure now upon them comes from non-renewable resources such as oil, sometimes to make synthetic clothing which is advocated as a humane substitute for fur. Similarly, in the case of the beaver, hunting and fur trade were thought to bring about the animal's demise, when in fact they were an integral part of its conservation. For many years children's books stated and still do, that the decline in the beaver population was due to the fur trade. In reality however, the decline in beaver numbers was because of habitat destruction and deforestation, as well as its continued persecution as a pest (it causes flooding). In Cree lands however, where the population valued the animal for meat and fur, it continued to thrive. The Inuit defend their relationship with the seal in response to outside critics.

Latin America (Bolivia)

The Izoceño-Guaraní of Santa Cruz Department, Bolivia is a tribe of hunters who were influential in establishing the Capitania del Alto y Bajo Isoso (CABI). CABI promotes economic growth and survival of the Izoceno people while discouraging the rapid destruction of habitat within Bolivia's Gran Chaco. They are responsible for the creation of the 34,000 square kilometre Kaa-Iya del Gran Chaco National Park and Integrated Management Area (KINP). The KINP protects the most biodiverse portion of the Gran Chaco, an ecoregion shared with Argentina, Paraguay and Brazil. In 1996, the Wildlife Conservation Society joined forces with CABI to institute wildlife and hunting monitoring programs in 23 Izoceño communities. The partnership combines traditional beliefs and local knowledge with the political and administrative tools needed to effectively manage habitats. The programs rely solely on voluntary participation by local hunters who perform self-monitoring techniques and keep records of their hunts. The information obtained by the hunters participating in the program has provided CABI with important data required to make educated decisions about the use of the land. Hunters have been willing participants in this program because of pride in their traditional activities, encouragement by their communities and expectations of benefits to the area.

Africa (Botswana)

In order to discourage illegal South African hunting parties and ensure future local use and sustainability, indigenous hunters in Botswana began lobbying for and implementing conservation practices in the 1960s. The Fauna Preservation Society of Ngamiland (FPS) was formed in 1962 by the husband and wife team: Robert Kay and June Kay, environmentalists working in conjunction with the Batawana tribes to preserve wildlife habitat.

The FPS promotes habitat conservation and provides local education for preservation of wildlife. Conservation initiatives were met with strong opposition from the Botswana government because of the monies tied to big-game hunting. In 1963, BaTawanga Chiefs and tribal hunter/adventurers in conjunction with the FPS founded Moremi National Park and Wildlife Refuge, the first area to be set aside by tribal people rather than governmental forces. Moremi National Park is home to a variety of wildlife, including lions, giraffes, elephants, buffalo, zebra, cheetahs and antelope, and covers an area of 3,000 square kilometers. Most of the groups involved with establishing this protected land were involved with hunting and were motivated by their personal observations of declining wildlife and habitat.

References

- Boleky, Vaughan (2005–2006). "Why Are Bat Houses Important?". Organization for Bat Conservation. Archived from the original on 2007-11-10. Retrieved 2007-11-17

- Meyer, Carrie A. (1993). "Environmental NGOs in Ecuador: An Economic Analysis of Institutional Change". The Journal of Developing Areas. 27 (2): 191–210. JSTOR 4192201

- Usher, M. B. (1986). Wildlife conservation evaluation: attributes, criteria and values. London, New York: Chapman and Hall. ISBN 978-94-010-8315-7

- Bélange, Claude (2004). "The Significance of the Eagle to the Indians". The Quebec History Encyclopedia. Marianopolis College. Retrieved 14 July 2012

- Daniel K. Rosenberg; Barry R. Noon; E. Charles Meslow (November 1997). "Biological Corridors: Form, Function, and Efficacy". BioScience. 47: 677–687. JSTOR view/1313208. doi:10.2307/1313208

- Klem, Jr., Daniel (1990). "Collisions between birds and windows: mortality and prevention." (PDF). Journal of Field Ornithology. 61 (l): 120–128. Archived from the original (PDF) on 2014-07-14

- Pevsner, Nikolaus (1961). The Buildings of England – Northamptonshire. London and New Haven: Yale University Press. pp. 94–5. ISBN 978-0-300-09632-3

- "Wildlife Conservation Efforts Are Violating Tribal Peoples' Rights". Deep Green Resistance News Service. Retrieved 10 June 2015

- Wilcove, David S; Wikelski, Martin (2008). "Going, Going, Gone: Is Animal Migration Disappearing". PLoS Biology. 6 (7): e188. PMC 2486312. PMID 18666834. doi:10.1371/journal.pbio.0060188

- Kala, Chandra Prakash (2009). "Medicinal plants conservation and enterprise development". Medicinal Plants - International Journal of Phytomedicines and Related Industries. 1 (2): 79–95. doi:10.5958/j.0975-4261.1.2.011

- Ehrlich, Anne H.; Ehrlich, Paul R. (1981). Extinction: the causes and consequences of the disappearance of species. New York: Random House. ISBN 0-394-51312-6

- Allen Best (November 1, 2010). "Wildlife and Highways: New Ideas Sought for Colorado's 'Berlin Wall'". New West. Retrieved March 3, 2013

- Harrison, J.R., Roberts, D.L., Hernandez-Castro, J. (2016). "Assessing the extent and nature of wildlife trade on the dark web". Conservation Biology. 30 (4): 900–904. doi:10.1111/cobi.12707

- McCallum, M (2007). "Amphibian Decline or Extinction? Current Declines Dwarf Background Extinction Rate". Journal of Herpetology. 41 (3): 483–491. doi:10.1670/0022-1511(2007)41[483:ADOECD]2.0.CO;2

- Wyman, Richard L. (1991). Global climate change and life on earth. New York: Routledge, Chapman and Hall. ISBN 0-412-02821-2

- Vié, J. C.; Hilton-Taylor, C.; Stuart, S.N. (2009). "Wildlife in a Changing World – An Analysis of the 2008 IUCN Red List of Threatened Species" (PDF). Gland, Switzerland: IUCN: 180. Retrieved December 24, 2010"

- Soulé, Michael E. (1986). "What is Conservation Biology?" (PDF). BioScience. American Institute of Biological Sciences. 35 (11): 727–34. JSTOR 1310054. doi:10.2307/1310054

An Overview of Marine Conservation

Marine conservation is the conservation of seas and oceans and the organisms found in them. The aim of this practice is to restrict the damage caused by humans to marine ecosystems. The topics discussed in the chapter are of great importance to broaden the existing knowledge on marine conservation.

Marine Conservation

Coral reefs have a great amount of biodiversity

Marine conservation is the protection and preservation of ecosystems in oceans and seas. Marine conservation focuses on limiting human-caused damage to marine ecosystems, restoring damaged marine ecosystems, and preserving vulnerable species of the marine life.

Overview

Marine conservation is a response to biological issues such as extinction and marine habitats change. Marine conservation is the study of conserving physical and biological marine resources and ecosystem functions. This is a relatively new discipline. Marine conservationists rely on a combination of scientific principles derived from marine biology, oceanography, and fisheries science, as well as on human factors such as demand for marine resources and marine law, economics and policy in order to determine how to best protect and conserve marine species and ecosystems. Marine conservation can be seen as sub-discipline of conservation biology.

Coral Reefs

Coral reefs are the epicenter of immense amounts of biodiversity, and are a key player in the survival of an entire ecosystem. They provide various marine animals with food, protection, and shelter which keep generations of species alive. Furthermore, coral reefs are an integral part of sustaining human life through serving as a food source (i.e., fish and mollusks) as well as a marine space for ecotourism which provides economic benefits. Also, humans are now conducting research regarding the use of corals as new potential sources for pharmaceuticals (i.e. steroids and anti-inflammatory drugs).

Unfortunately, because of the human impact on coral reefs, these ecosystems are becoming increasingly degraded and in need of conservation. The biggest threats include overfishing, destructive fishing practices, and sedimentation and pollution from land-based sources. This, in conjunction with increased carbon in oceans, coral bleaching, and diseases, means that there are no pristine reefs anywhere in the world. Up to 88% of coral reefs in Southeast Asia are now threatened, with 50% of those reefs at either "high" or "very high" risk of disappearing, which directly affects the biodiversity and survival of species dependent on coral.

This is especially harmful to island nations such as Samoa, Indonesia, and the Philippines, because many people there depend on the coral reef ecosystems to feed their families and to make a living. However, many fishermen are unable to catch as many fish as they used to, so they are increasingly using cyanide and dynamite in fishing, which further degrades the coral reef ecosystem. This perpetuation of bad habits simply leads to the further decline of coral reefs and therefore perpetuates the problem. One way of stopping this cycle is by educating the local community about why the conservation of marine spaces that include coral reefs is important. Once the local communities understand the personal stakes, then they will fight to preserve the reefs. Conserving coral reefs has many economic, social, and ecological benefits, not only for the people who live on these islands, but for people throughout the world.

Human Impact

The deterioration of coral reefs is mainly linked to human activities – 88% of reefs are threatened through various reasons as listed above, including excessive amounts of CO_2 (carbon dioxide) emissions. Oceans absorb approximately 1/3 of the CO_2 produced by humans, which has detrimental effects on the marine environment. The increasing levels of CO_2 in oceans change the seawater chemistry by decreasing the pH. This process is also known as ocean acidification. Acidification negatively affects the carbonate buffering system and drops the carbonate saturation by 30%, which results in a decrease in reef calcification. Reductions in calcification have negative implications on calcifiers, such as corals and shellfish. Some examples include diminishing coral resilience from bleaching, decreasing organisms' ability to fight off predators, inhibiting their potential to compete for food, and altering behavior patterns. When the bottom of the food web declines tremendously due to acidification, the food web and the whole marine conservation effort is jeopardized. Although humans cause the greatest threat to the marine environment, they also have the ability to create effective management plans that will be the key to successful marine conservation. Although the most widely known conservation tool is the MPA, one of the best marine conservation tools simply stems from smarter individual choices made in efforts to reduce CO_2 emissions on a daily basis.

Techniques

Strategies and techniques for marine conservation tend to combine theoretical disciplines, such as population biology, with practical conservation strategies, such as setting up protected areas, as with marine protected areas (MPAs) or Voluntary Marine Conservation Areas. Other techniques include developing sustainable fisheries and restoring the populations of endangered species through artificial means.

Another focus of conservationists is on curtailing human activities that are detrimental to either marine ecosystems or species through policy, techniques such as fishing quotas, like those set up by the Northwest Atlantic Fisheries Organization, or laws such as those listed below. Recognizing the economics involved in human use of marine ecosystems is key, as is education of the public about conservation issues. This includes educating tourists that come to an area who might not be familiar with certain regulations regarding the marine habitat. One example of this is a project called Green Fins based in Southeast Asia that uses the scuba diving industry to educate the public. This project, implemented by UNEP, encourages scuba diving operators to educate their students about the importance of marine conservation and encourage them to dive in an environmentally friendly manner that does not damage coral reefs or associated marine ecosystems.

Technology and Halfway Technology

Marine conservation technologies are used to protect endangered and threatened marine organisms and/or habitat. These technologies are innovative and revolutionary because they reduce bycatch, increase the survivorship and health of marine life and habitat, and benefit fishermen who depend on the resources for profit. Examples of technologies include marine protected areas (MPAs), turtle excluder devices (TEDs), autonomous recording unit, pop-up satellite archival tag, and radio-frequency identification (RFID). Commercial practicality plays an important role in the success of marine conservation because it is necessary to cater to the needs of fishermen while also protecting marine life.

Pop-up satellite archival tag (PSAT or PAT) plays a vital role in marine conservation by providing marine biologists with an opportunity to study animals in their natural environments. These are used to track movements of (usually large, migratory) marine animals. A PSAT is an archival tag (or data logger) that is equipped with a means to transmit the collected data via satellite. Though the data are physically stored on the tag, its major advantage is that it does not have to be physically retrieved like an archival tag for the data to be available, making it a viable independent tool for animal behavior studies. These tags have been used to track movements of ocean sunfish, marlin, blue sharks, bluefin tuna, swordfish and sea turtles. Location, depth, temperature, and body movement data are used to answer questions about migratory patterns, seasonal feeding movements, daily habits, and survival after catch and release.

Turtle excluder devices (TEDs) remove a major threat to turtles in their marine environment. Many sea turtles are accidentally captured, injured or killed by fishing. In response to this threat the National Oceanic and Atmospheric Administration (NOAA) worked with the shrimp trawling industry to create the TEDs. By working with the industry they insured the commercial viability of the devices. A TED is a series of bars that is placed at the top or bottom of a trawl net, fitting the bars into the "neck" of the shrimp trawl and acting as a filter to ensure that only small animals may pass through. The shrimp will be caught but larger animals such as marine turtles that become caught by the trawler will be rejected by the filter function of the bars.

Similarly, halfway technologies work to increase the population of marine organisms. However, they do so without behavioral changes, and address the symptoms but not the cause of the declines. Examples of halfway technologies include hatcheries and fish ladders.

Laws and Treaties

International laws and treaties related to marine conservation include the 1966 Convention on Fishing and Conservation of Living Resources of the High Seas. United States laws related to marine conservation include the 1972 Marine Mammal Protection Act, as well as the 1972 Marine Protection, Research and Sanctuaries Act, which established the National Marine Sanctuaries program.

In 2010, the Scottish Parliament enacted new legislation for the protection of marine life with the Marine (Scotland) Act 2010. Its provisions include marine planning, marine licensing, marine conservation, seal conservation, and enforcement.

Organizations and Education

The shore of the Pacific Ocean in San Francisco, California.

There are marine conservation organizations throughout the world that focus on funding conservation efforts, educating the public and stakeholders, and lobbying for conservation law and policy. Examples of these include Oceana, the Marine Conservation Institute (United States), Blue Frontier Campaign (United States), Sea Shepherd Conservation Society (international), Frontier (the Society for Environmental Exploration) (United Kingdom), Marine Conservation Society (United Kingdom), Community Centred Conservation (C3), the Reef-World Foundation (United Kingdom), Reef Watch (India), and Australian Marine Conservation Society. Zoox (United Kingdom) is an example of an organization that provides both marine conservation training and professional career development to volunteers who are also working on marine conservation projects such as Green Fins.

On a regional level, PERSGA, the Regional Organization for the Conservation of the Environment of the Red Sea and the Gulf of Aden, is a regional entity which serves as the secretariat for the Jeddah Convention-1982, one of the first regional marine agreements. PERSGA member states are Djibouti, Egypt, Jordan, Saudi Arabia, Somalia, Sudan and Yemen.

Extinct and Endangered Species

Marine Mammals

Baleen whales were predominantly hunted from 1600 through the mid-1900s, and were nearing extinction when a global ban on commercial whaling was put into effect in 1896 by the IWC (International Whaling Convention). The Atlantic gray whale, last sighted in 1740, is now extinct due to European and Native American Whaling. Since the 1960s the global population of monk seals has been rapidly declining. The Hawaiian and Mediterranean monk seals are considered to be one of the most endangered marine mammals on the planet, according to the NOAA. The last sighting of the Caribbean monk seal was in 1952, and it has now been confirmed extinct by the NOAA. The vaquita porpoise, discovered in 1958, has become the most endangered marine species. Over half the population has disappeared since 2012, leaving 100 left in 2014. The vaquita frequently drowns in fishing nets, which are used illegally in marine protected areas off the Gulf of Mexico.

Sea Turtles

In 2004, the Marine Turtle Specialist Group (MTSG), from the International Union for Conservation of Nature (IUCN), ran an assessment which determined that green turtles were globally endangered. Population decline in ocean basins is indicated through data collected by the MTSG that analyzes abundance and historical information on the species. This data examined the global population of green turtles at 32 nesting sites, and determined that over the last 100–150 years there has been a 48–65 percent decrease in the number of mature nesting females. The Kemp's ridley sea turtle population fell in 1947 when 33,000 nests, which accounted for 80 percent of the population, were collected and sold by villagers in Racho Nuevo, Mexico. In the early 1960s only 5,000 individuals were left, and between 1978 and 1991, 200 Kemp's ridley turtles nested annually. In 2015, the World Wildlife Fund and *National Geographic Magazine* named the Kemp's ridley the most endangered sea turtle in the world, with 1000 females nesting annually.

Fish

In 2014, the IUCN moved the Pacific bluefin tuna from "least concerned" to "vulnerable" on a scale that represents level of extinction risk. The Pacific bluefin tuna is targeted by the fishing industry mainly for its use in sushi. A stock assessment released in 2013 by the International Scientific Committee for Tuna and Tuna-Like Species in the North Pacific Ocean (ISC) shows that the Pacific bluefin tuna population dropped by 96 percent in the Pacific Ocean. According to the ISC assessment, 90 percent of the Pacific bluefin tuna caught are juveniles that have not reproduced.

Between 2011 and 2014, the European eel, Japanese eel, and American eel were put on the IUCN red list of endangered species. In 2015, the Environmental Agency concluded that the number of European eels has declined by 95 percent since 1990. An Environmental Agency officer, Andy Don, who has been researching eels for the past 20 years, said, "There is no doubt that there is a crisis.

People have been reporting catching a kilo of glass eels this year when they would expect to catch 40 kilos. We have got to do something."

Marine Plants

Johnson's seagrass, a food source for the endangered green sea turtle, is the scarcest species in its genus. It reproduces asexually, which limits its ability to populate and colonize habitats. Data on this species is limited, but it is known that since the 1970s there has been a 50 percent decrease in abundance.

History of Marine Conservation

Modern marine conservation first became globally recognized in the 1970s after World War II in an era known as the "marine revolution". The United States federal legislation showed its support of marine conservation by institutionalizing protected areas and creating marine estuaries. In the mid-1970s the United States formed the International Union for Conservation of Nature (IUCN). Through this program, nations could communicate and make agreements about marine conservation. After the formation of the IUCN, new independent organizations known as non-governmental organizations started to appear. These organizations were self-governed and had individual goals for marine conservation. At the end of the 1970s, undersea explorations equipped with new technology such as computers were undertaken. During these explorations, fundamental principles of change were discovered in relation to marine ecosystems. Through this discovery, the interdependent nature of the ocean was revealed. This led to a change in the approach of marine conservation efforts, and a new emphasis was put on restoring systems within the environment, along with protecting biodiversity.

Overabundance

Overabundance occurs when the population of a certain species cannot be controlled. The domination of one species can create an imbalance in an ecosystem, which can lead to the demise of other species and of the habitat. Overabundance occurs predominately in invasive species. Cargo ships introduce new species into different environments through releasing ballast water into an ecosystem. A tank of ballast water is estimated to contain around 3,000 non-native species.

The San Francisco Bay is one of the places in the world that is the most impacted by foreign and invasive species. According to the Baykeeper organization, 97 percent of the organisms in the San Francisco Bay have been compromised by the 240 invasive species that have been brought into the ecosystem. Invasive species in the bay such as the Asian clam have changed the food web of the ecosystem by depleting populations of native species such as plankton. The Asian clam clogs pipes and obstructs the flow of water in electrical generating facilities. Their presence in the San Francisco Bay has cost the United States an estimated one billion dollars in damages.

Marine Protected Area

Marine protected areas (MPA) are protected areas of seas, oceans, estuaries or large lakes. MPAs restrict human activity for a conservation purpose, typically to protect natural or cultural resources. Such marine resources are protected by local, state, territorial, native, regional, national, or international authorities and differ substantially among and between nations. This variation includes different limitations on development, fishing practices, fishing seasons and catch limits, moorings and bans on removing or disrupting marine life. In some situations (such as with the Phoenix Islands Protected Area), MPAs also provide revenue for countries, potentially equal to the income that they would have if they were to grant companies permissions to fish.

Milford Sound, New Zealand is a *strict marine reserve* (Category Ia)
Mitre Peak, the mountain at left, rises 1,692 m (5,551 ft) above the sea.

On 28 October 2016 in Hobart, Australia, the Convention for the Conservation of Antarctic Marine Living Resources agreed to establish the first Antarctic and largest marine park in the world encompassing 1.55 million km² (600,000 sq mi) in the Ross Sea. Other large MPAs are in the Indian, Pacific, and Atlantic Oceans in certain exclusive economic zones of Australia and overseas territories of France, the United Kingdom and the United States, with major (990,000 square kilometres (380,000 sq mi) or larger) new or expanded MPAs by these nations since 2012—such as Natural Park of the Coral Sea, Pacific Remote Islands Marine National Monument, Coral Sea Commonwealth Marine Reserve and South Georgia and the South Sandwich Islands Marine Protected Area. When counted with MPAs of all sizes from many other countries, as of August 2016 there are more than 13,650 MPAs, encompassing 2.07% of the world's oceans, with half of that area – encompassing 1.03% of the world's oceans – receiving complete "no-take" designation.

Terminology

"MPA" is an umbrella term for protected areas that includes some area of marine landscape and/or biodiversity. The IUCN defines a marine protected area as:

"Any area of the intertidal or subtidal terrain, together with its overlying water and associated flora, fauna, historical and cultural features, which has been reserved by law or other effective means to protect part or all of the enclosed environment."

An alternative is "a clearly defined geographical space, recognized, dedicated, and managed through legal or other effective means, to achieve the long term conservation of nature with associated ecosystem services and cultural values". United States Executive Order 13158 in May 2000 established MPAs, defining them as;

"Any area of the marine environment that has been reserved by federal, state, tribal, territorial, or local laws or regulations to provide lasting protection for part or all of the natural and cultural resources therein."

The Convention on Biological Diversity defined the broader term of *marine and coastal protected area* (MCPA);

"Any defined area within or adjacent to the marine environment, together with its overlying water and associated flora, fauna, historical and cultural features, which has been reserved by legislation or other effective means, including custom, with the effect that its marine and/or coastal biodiversity enjoys a higher level of protection than its surroundings."

Classifications

The Chagos Archipelago was declared the world's largest marine reserve in April 2010 with an area of 250,000 square miles until March 2015 when It was declared illegal by the Permanent Court of Arbitration.

Several types of compliant MPA can be distinguished:

- A totally marine area with no significant terrestrial parts.

- An area containing both marine and terrestrial components, which can vary between two extremes; those that are predominantly maritime with little land (for example, an atoll would have a tiny island with a significant maritime population surrounding it), or that is mostly terrestrial.

- Marine ecosystems that contain land and intertidal components only. For example, a mangrove forest would contain no open sea or ocean marine environment, but its river-like marine ecosystem nevertheless complies with the definition.

IUCN offered seven categories of protected area, based on management objectives and four broad governance types.

Cat	IUCN Protected Area Management Categories:
Ia	Strict nature reserve A marine reserve usually connotes "maximum protection", where all resource removals are strictly prohibited. In countries such as Kenya and Belize, marine reserves allow for low-risk removals to sustain local communities.

Ib	Wilderness area
II	National park Marine parks emphasize the protection of ecosystems but allow light human use. A marine park may prohibit fishing or extraction of resources, but allow recreation. Some marine parks, such as those in Tanzania, are zoned and allow activities such as fishing only in low risk areas.
III	Natural monuments or features Established to protect historical sites such as shipwrecks and cultural sites such as aboriginal fishing grounds.
IV	Habitat/species management area Established to protect a certain species, to benefit fisheries, rare habitat, as spawning/nursing grounds for fish, or to protect entire ecosystems.
V	Protected seascape Limited active management, as with protected landscapes.
VI	Sustainable use of natural resources

Related protected area categories include the following;

- World Heritage Site (WHS) – an area exhibiting extensive natural or cultural history. Maritime areas are poorly represented, however, with only 46 out of over 800 sites.

- Man and the Biosphere – UNESCO program that promotes "a balanced relationship between humans and the biosphere". Under article 4, biosphere reserves must "encompass a mosaic of ecological systems", and thus combine terrestrial, coastal, or marine ecosystems. In structure they are similar to Multiple-use MPAs, with a core area ringed by different degrees of protection.

- Ramsar site – must meet certain criteria for the definition of "Wetland" to become part of a global system. These sites do not necessarily receive protection, but are indexed by importance for later recommendation to an agency that could designate it a protected area.

While "area" refers to a single contiguous location, terms such as *"network"*, *"system"*, and *"region"* that group MPAs are not always consistently employed."*System*" is more often used to refer to an individual MPA, whereas *"region"* is defined by the World Conservation Monitoring Centre as:

"A collection of individual MPAs operating cooperatively, at various spatial scales and with a range of protection levels that are designed to meet objectives that a single reserve cannot achieve."

At the 2004 Convention on Biological Diversity, the agency agreed to use *"network"* on a global level, while adopting *system* for national and regional levels. The *network* is a mechanism to establish regional and local systems, but carries no authority or mandate, leaving all activity within the *"system"*.

No take zones (NTZs), are areas designated in a number of the world's MPAs, where all forms of exploitation are prohibited and severely limits human activities. These no take zones can cover an entire MPA, or specific portions. For example, the 1,150,000 square kilometres (440,000 sq mi) Papahānaumokuākea Marine National Monument, the world's largest MPA (and largest protected area of any type, land or sea), is a 100% no take zone.

Related terms include; *specially protected area* (SPA), *Special Area of Conservation* (SAC), the United Kingdom's *marine conservation zones* (MCZs), or *area of special conservation* (ASC) etc. which each provide specific restrictions.

Stressors

Stressors that affect oceans include "the impact of extractive industries, localised pollution, and changes to its chemistry (ocean acidification) resulting from elevated carbon dioxide levels, due to our emissions". MPAs have been cited as the ocean's single greatest hope for increasing the resilience of the marine environment to such stressors. Well-designed and managed MPAs developed with input and support from interested stakeholders can conserve biodiversity and protect and restore fisheries.

Economics

MPAs can help sustain local economies by supporting fisheries and tourism. For example, Apo Island in the Philippines made protected one quarter of their reef, allowing fish to recover, jump-starting their economy. This was shown in the film, *Resources at Risk: Philippine Coral Reef*. A 2016 report by the Center for Development and Strategy found that programs like the United States National Marine Sanctuary system can develop considerable economic benefits for communities through Public–private partnerships.

Management

Typical MPAs restrict fishing, oil and gas mining and/or tourism. Other restrictions may limit the use of ultrasonic devices like sonar (which may confuse the guidance system of cetaceans), development, construction and the like. Some fishing restrictions include "no-take" zones, which means that no fishing is allowed. Less than 1% of US MPAs are no-take.

Ship transit can also be restricted or banned, either as a preventive measure or to avoid direct disturbance to individual species. The degree to which environmental regulations affect shipping varies according to whether MPAs are located in territorial waters, exclusive economic zones, or the high seas. The law of the sea regulates these limits.

Most MPAs have been located in territorial waters, where the appropriate government can enforce them. However, MPAs have been established in exclusive economic zones and in international waters. For example, Italy, France and Monaco in 1999 jointly established a cetacean sanctuary in the Ligurian Sea named the Pelagos Sanctuary for Mediterranean Marine Mammals. This sanctuary includes both national and international waters. Both the CBD and IUCN recommended a variety of management systems for use in a protected area system. They advocated that MPAs be seen as one of many "nodes" in a network of protected areas. The following are the most common management systems:

Seasonal and temporary management—Activities, most critically fishing, are restricted seasonally or temporarily, e.g., to protect spawning/nursing grounds or to let a rapidly reducing species recover.

Multiple-use MPAs—These are the most common and arguably the most effective. These areas employ two or more protections. The most important sections get the highest protection, such as a no take zone and are surrounded with areas of lesser protections.

Community involvement and related approaches—Community-managed MPAs empower local communities to operate partially or completely independent of the governmental jurisdictions they occupy. Empowering communities to manage resources can lower conflict levels and enlist the support of diverse groups that rely on the resource such as subsistence and commercial fishers, scientists, recreation, tourism businesses, youths and others.

Asinara, Italy is listed by WDPA as both a marine reserve and a national marine park, and as such could be labelled 'multiple-use'.

MPA networks—"A group of MPAs that interact with one another ecologically and/or socially form a network". These networks are intended to connect individuals and MPAs and promote education and cooperation among various administrations and user groups. "MPA networks are, from the perspective of resource users, intended to address both environmental and socio-economic needs, complementary ecological and social goals and designs need greater research and policy support". Filipino communities connect with one another to share information about MPAs, creating a larger network through the social communities' support. Emerging or established MPA networks can be found in Australia, Belize, the Red Sea, Gulf of Aden and Mexico.

International Efforts

The 17th International Union for Conservation of Nature (IUCN) General Assembly in San Jose, California, the 19th IUCN assembly and the fourth World Parks Congress all proposed to centralise the establishment of protected areas. The World Summit on Sustainable Development in 2002 called for

the establishment of marine protected areas consistent with international laws and based on scientific information, including representative networks by 2012.

The Evian agreement, signed by G8 Nations in 2003, agreed to these terms. The Durban Action Plan, developed in 2003, called for regional action and targets to establish a network of protected areas by 2010 within the jurisdiction of regional environmental protocols.It recommended establishing protected areas for 20 to 30% of the world's oceans by the goal date of 2012. The Convention on Biological Diversity considered these recommendations and recommended requiring countries to set up marine parks controlled by a central organization before merging them. The United Nations Framework Convention on Climate Change agreed to the terms laid out by the convention, and in 2004, its member nations committed to the following targets;

- By 2006 complete an area system gap analysis at national and regional levels.

- By 2008 address the less represented marine ecosystems, accounting for those beyond national jurisdiction in accordance.

- By 2009 designate the protected areas identified through the gap analysis.

- By 2012 complete the establishment of a comprehensive and ecologically representative network.

Bunaken Marine Park, Indonesia is officially listed as both a marine reserve and a national marine park.

"The establishment by 2010 of terrestrial and by 2012 for marine areas of comprehensive, effectively managed, and ecologically representative national and regional systems of protected areas that collectively, inter alia through a global network, contribute to achieving the three objectives of the Convention and the 2010 target to significantly reduce the current late of biodiversity loss at the global, regional, national, and sub-national levels and contribute to poverty reduction and the pursuit of sustainable development."

The UN later endorsed another decision, Decision VII/15, in 2006:

Effective conservation of 10% of each of the world's ecological regions by 2010.

— United Nations Framework Convention on Climate Change Decision VII/15

United Nations Convention on the Law of The Sea

The Antarctic Treaty System

On 7 April 1982, the Convention on the Conservation of Antarctic Marine Living Resources (CAMLR Convention) came into force after discussions began in 1975 between parties of the then-current Antarctic Treaty to limit large-scale exploitation of krill by commercial fisheries. The Convention bound contracting nations to abide by previously agreed upon Antarctic territorial claims and peaceful use of the region while protecting ecosystem integrity south of the Antarctic Convergence and 60 S latitude. In so doing, it also established a commission of the original signatories and acceding parties called the Commission for the Conservation of Antarctic Marine Living Resources (CCAMLR) to advance these aims through protection, scientific study, and rational use, such as harvesting, of those marine resources. Though separate, the Antarctic Treaty and CCAMLR, make up part the broader system of international agreements called the Antarctic Treaty System. Since 1982, the CCAMLR meets annually to implement binding conservations measures like the creation of 'protected areas' at the suggestion of the convention's scientific committee.

In 2009, the CCAMLR created the first 'high-seas' MPA entirely within international waters over the southern shelf of the South Orkney Islands. This area encompasses 94,000 square kilometres (36,000 sq mi) and all fishing activity including transhipment, and dumping or discharge of waste is prohibited with the exception of scientific research endeavors. On 28 October 2016, the CCAMLR, composed of 24 member countries and the European Union at the time, agreed to establish the world's largest marine park encompassing 1.55 million km² (600,000 sq mi) in the Ross Sea after several years of failed negotiations. Establishment of the Ross Sea MPA required unanimity of the commission members and enforcement will begin in December 2017. However, due to a sunset provision inserted into the proposal, the new marine park will only be in force for 35 years.

National Targets

Many countries have established national targets, accompanied by action plans and implementations. The UN Council identified the need for countries to collaborate with each other to establish effective regional conservation plans. Some national targets are listed in the table below:

Country	Plan of action
American Samoa	20% of reefs to be protected by 2010
Australia – South Australia	19 marine protected areas by 2010
Bahamas	20% of the marine ecosystem protected for fishery replenishment by 2010. 20% of coastal and marine habitats by 2015.
Belize	20% of bioregions. 30% of Coral reefs. 60% of turtle nesting sites. 30% of Manatee distribution. 60% of American crocodile nesting. 80% of breeding areas.
Chile	10% of marine areas by 2010. National network for organization by 2015.

Cuba	22% of land habitat, including:
	     15% of the insular shelf
	     25% of coral reefs
	     25% of wetlands

Dominican Republic	20% of marine and coastal by 2020.
Micronesia	30% of shoreline ecosystems by 2020.

Fiji	30% of reefs by 2015.
	30% of water managed by marine protected areas by 2020.
Germany	38% of water managed by the marine protected network. (no set date)
Grenada	25% of nearby marine resources by 2020.
Guam	30% of nearby marine ecosystem by 2020.
Indonesia	100,000 km^2 by 2010.
	200,000 km^2 by 2020.
Ireland	14% of territorial waters as of 2009
Jamaica	20% of marine habitats by 2020.
Madagascar	100,000 km^2 by 2012.
Marshal Islands	30% of nearby marine ecosystem by 2020.
New Zealand	20% of marine environment by 2010.
North Mariana Islands	30% of nearby marine ecosystem by 2020.
Palau	30% of nearby marine ecosystem by 2020.
Peru	Marine protected area system established by 2015.
Philippines	10% fully protected by 2020.
Senegal	Creation of MPA network. (no set date)
St. Vincent and the Grenadines	20% of marine areas by 2020.
Tanzania	10% of marine area by 2010; 20% by 2020.
United Kingdom	Establish an ecologically coherent network of marine protected areas by 2012.
United States – California	29 MPAs covering 18% of state marine area with 243 square kilometres (94 sq mi) at maximum protection.

National Efforts

The marine protected area network is still in its infancy. As of October 2010, approximately 6,800 MPAs had been established, covering 1.17% of global ocean area. Protected areas covered 2.86% of exclusive economic zones (EEZs). MPAs covered 6.3% of territorial seas. Many prohibit the use of harmful fishing techniques yet only 0.01% of the ocean's area is designated as a "no take zone". This coverage is far below the projected goal of 20%-30% Those targets have been questioned mainly due to the cost of managing protected areas and the conflict that protections have generated with human demand for marine goods and services.

Greater Caribbean

The Greater Caribbean subdivision encompasses an area of about 5,700,000 square kilometres (2,200,000 sq mi) of ocean and 38 nations. The area includes island countries like the Bahamas and Cuba, and the majority of Central America. The Convention for Protection and Development of the Marine Environment of the Wider Caribbean Region (better known as the Cartagena Convention) was established in 1983. Protocols involving protected areas were ratified in 1990. As of 2008, the region hosted about 500 MPAs. Coral reefs are the best represented.

The Caribbean region; the UNEP–defined region also includes the Gulf of Mexico. This region is encompassed by the Mesoamerican Barrier Reef System proposal, and the Caribbean challenge.

The Gulf of Mexico region (in 3D) is encompassed by the "Islands in the Stream" proposal.

Two networks are under development, the Mesoamerican Barrier Reef System (a long barrier reef that borders the coast of much of Central America), and the "Islands in the Stream" program (covering the Gulf of Mexico).

Southeast Asia

Southeast Asia is a global epicenter for marine diversity. 12% of its coral reefs are in MPAs. The Philippines have some the world's best coral reefs and protect them to attract international tourism. Most of the Philippines' MPAs are established to secure protection for its coral reef and sea grass habitats. Indonesia has MPAs designed for tourism and relies on tourism as a main source of income.

Philippines

The Philippines host one of the most highly biodiverse regions, with 464 reef-building coral species. Due to overfishing, destructive fishing techniques, and rapid coastal development, these are in rapid decline. The country has established some 600 MPAs. However, the majority are poorly enforced and are highly ineffective. However, some have positively impacted reef health, increased fish biomass, decreased coral bleaching and increased yields in adjacent fisheries. One notable example is the MPA surrounding Apo Island.

Latin America

Latin America has designated one large MPA system. As of 2008, 0.5% of its marine environment was protected, mostly through the use of small, multiple-use MPAs.

South Pacific

The South Pacific network ranges from Belize to Chile. Governments in the region adopted the Lima convention and action plan in 1981. An MPA-specific protocol was ratified in 1989. The permanent commission on the exploitation and conservation on the marine resources of the South Pacific promotes the exchange of studies and information among participants.

The region is currently running one comprehensive cross-national program, the Tropical Eastern Pacific Marine Corridor Network, signed in April 2004. The network covers about 211,000,000 square kilometres (81,000,000 sq mi).

One alternative to imposing MPAs on an indigenous population is through the use of Indigenous Protected Areas, such as those in Australia.

North Pacific

The North Pacific network covers the western coasts of Mexico, Canada, and the U.S. The "Antigua Convention" and an action plan for the north Pacific region were adapted in 2002. Participant nations manage their own national systems. In 2010-2011, the State of California completed hearings and actions via the state Department of Fish and Game to establish new MPAs.

United States and Pacific Island Territories

President Barack Obama signed a proclamation on September 25, 2014, designating the world's largest marine reserve. The proclamation expanded the existing Pacific Remote Islands Marine National Monument, one of the world's most pristine tropical marine environments, to six times its current size, encompassing 490,000 square miles (1,300,000 km^2) of protected area around these islands. Expanding the Monument protected the area's unique deep coral reefs and seamounts.

In April 2009, the US established a United States National System of Marine Protected Areas, which strengthens the protection of US ocean, coastal and Great Lakes resources. These large-scale MPAs should balance "the interests of conservationists, fishers, and the public." As of 2009, 225 MPAs participated in the national system. Sites work together toward common national and regional conservation goals and priorities. NOAA's national marine protected areas center maintains a comprehensive inventory of all 1,600+ MPAs within the US exclusive economic zone. Most US MPAs.allow some type of extractive use. Fewer than 1% of U.S. waters prohibit all extractive activities.

Diagram illustrating the orientation of the 3 marine sanctuaries of Central California: Cordell Bank, Gulf of the Farallones, and Monterey Bay. Davidson Seamount, part of the Monterey Bay sanctuary, is indicated at bottom-right.

In 1981 Olympic National Park became a marine protected area. The total protected site area is 3,697 square kilometres (1,427 sq mi). 173.2 km² of the area was an MPA. The national system is a mechanism to foster MPA collaboration. Sites that meet pertinent criteria are eligible to join the national system. Four entry criteria govern admission:

- Meets the definition of an MPA as defined in the Framework.

- Has a management plan (can be sitespecific or part of a broader programmatic management plan; must have goals and objectives and call for monitoring or evaluation of those goals and objectives).

- Contributes to at least one priority conservation objective as listed in the Framework.

- Cultural heritage MPAs must also conform to criteria for the National Register for Historic Places."

In 1999, California adopted the Marine Life Protection Act, establishing the first state law requiring a comprehensive, science-based MPA network. The state created the Marine Life Protection Act Initiative. The MLPA Blue Ribbon Task Force and stakeholder and scientific advisory groups ensure that the process uses the science and public participation.

The MLPA Initiative established a plan to create California's statewide MPA network by 2011 in several steps. The Central Coast step was successfully completed in September, 2007. The North Central Coast step was completed in 2010. The South Coast and North Coast steps were expected to go into effect in 2012.

United Kingdom and British Overseas Territories

United Kingdom

There are a number of marine protected areas around the coastline of the United Kingdom, known as Marine Conservation Zones in England, Wales, and Northern Ireland, Marine Protection Areas in Scotland. They are to be found in inshore and offshore waters.

British Overseas Territories

The United Kingdom is also creating marine protected reserves around several British Overseas Territories. The UK is responsible for 6.8 million square kilometres of ocean around the world, larger than all but four other countries.

The Chagos Marine Protected Area in the Indian Ocean was established in 2010 as a "no-take-zone". With a total surface area of 640,000 square kilometres (250,000 sq mi), it was the world's largest contiguous marine reserve. In March 2015, the UK announced the creation of a marine reserve around the Pitcairn Islands in the Southern Pacific Ocean to protect its special biodiversity. The area of 830,000 square kilometres (320,000 sq mi) surpassed the Chagos Marine Protected Area as the world's largest contiguous marine reserve, until the August 2016 expansion of the Papahānaumokuākea Marine National Monument in the United States to 1,510,000 square kilometres (580,000 sq mi).

In January 2016, the UK government announced the intention to create a marine protected area around Ascension Island. The protected area will be 234,291 square kilometres (90,460 sq mi), half of which will be closed to fishing.

Europe

The Natura 2000 ecological MPA network in the European Union included MPAs in the North Atlantic, the Mediterranean Sea and the Baltic Sea. The member states had to define NATURA 2000 areas at sea in their Exclusive Economic Zone.

Two assessments, conducted thirty years apart, of three Mediterranean MPAs, demonstrate that proper protection allows commercially valuable and slow-growing red coral (Corallium rubrum) to produce large colonies in shallow water of less than 50 metres (160 ft). Shallow-water colonies outside these decades-old MPAs are typically very small. The MPAs are Banyuls, Carry-le-Rouet and Scandola, off the island of Corsica.

- Mediterranean Science Commission; proposed the creation of 7 marine protected areas ("peace parcs")

Notable Marine Protected Areas

- The Bowie Seamount Marine Protected Area off the coast of British Columbia, Canada.
- The Great Barrier Reef Marine Park in Queensland, Australia.
- The Ligurian Sea Cetacean Sanctuary in the seas of Italy, Monaco and France.
- The Dry Tortugas National Park in the Florida Keys, USA.
- The Papahānaumokuākea Marine National Monument in Hawaii.
- The Phoenix Islands Protected Area, Kiribati.

- The Channel Islands National Marine Sanctuary in California, USA.

- The Chagos Marine Protected Area in the Indian Ocean.

- The Wadden Sea bordering the North Sea in the Netherlands, Germany, and Denmark.

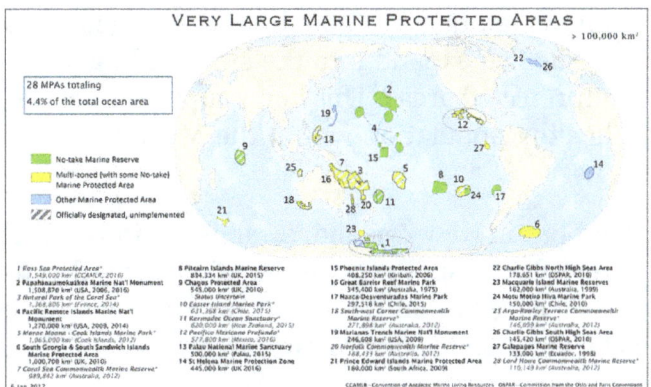

Marine protected areas

Assessment

The Prickly Pear Cays are a marine protected area, roughly six miles from Road Bay, Anguilla, in the Leeward Islands of the Caribbean.

Managers and scientists use geographic information systems and remote sensing to map and analyze MPAs. NOAA Coastal Services Center compiled an "Inventory of GIS-Based Decision-Support Tools for MPAs." The report focuses on GIS tools with the highest utility for MPA processes. Remote sensing uses advances in aerial photography image capture, pop-up archival satellite tags, satellite imagery, acoustic data, and radar imagery. Mathematical models that seek to reflect the complexity of the natural setting may assist in planning harvesting strategies and sustaining fishing grounds.

Coral Reefs

Coral reef systems have been in decline worldwide. Causes include overfishing, pollution and ocean acidification. As of 2013 30% of the world's reefs were severely damaged. Approximately 60% will be lost by 2030 without enhanced protection. Marine reserves with "no take zones" are the most effective form of protection. Only about 0.01% of the world's coral reefs are inside effective MPAs.

Fish

MPAs can be an effective tool to maintain fish populations. The general concept is to create over-population within the MPA. The fish expand into the surrounding areas to reduce crowding, increasing the population of unprotected areas. This helps support local fisheries in the surrounding area, while maintaining a healthy population within the MPA. Such MPAs are most commonly used for coral reef ecosystems.

One example is at Goat Island Bay in New Zealand, established in 1977. Research gathered at Goat Bay documented the spillover effect. "Spillover and larval export—the drifting of millions of eggs and larvae beyond the reserve—have become central concepts of marine conservation". This positively impacted commercial fishermen in surrounding areas.

Another unexpected result of MPAs is their impact on predatory marine species, which in some conditions can increase in population. When this occurs, prey populations decrease. One study showed that in 21 out of 39 cases, "trophic cascades," caused a decrease in herbivores, which led to an increase in the quantity of plant life. (This occurred in the Malindi Kisite and Watamu Marian National Parks in Kenya; the Leigh Marine Reserve in New Zealand; and Brackett's Landing Conservation Area in the US.

Success Criteria

Both CBD and IUCN have criteria for setting up and maintaining MPA networks, which emphasize 4 factors:

- Adequacy—ensuring that the sites have the size, shape, and distribution to ensure the success of selected species.

- Representability—protection for all of the local environment's biological processes

- Resilience—the resistance of the system to natural disaster, such as a tsunami or flood.

- Connectivity—maintaining population links across nearby MPAs.

Misconceptions

Misconceptions about MPAs include the belief that all MPAs are no-take or no-fishing areas. However, less than 1 percent of US waters are no-take areas. MPA activities can include consumption fishing, diving and other activities.

Another misconception is that most MPAs are federally managed. Instead, MPAs are managed under hundreds of laws and jurisdictions. They can be exist in state, commonwealth, territory and tribal waters.

Another misconception is that a federal mandate dedicates a set percentage of ocean to MPAs. Instead the mandate requires an evaluation of current MPAs and creates a public resource on current MPAs.

Criticism

Some existing and proposed MPAs have been criticized by indigenous populations and their supporters, as impinging on land usage rights. For example, the proposed Chagos Protected Area in the Chagos Islands is contested by Chagossians deported from their homeland in 1965 by the British as part of the creation of the British Indian Ocean Territory (BIOT). According to Wikileaks CableGate documents, the UK proposed that the BIOT become a "marine reserve" with the aim of preventing the former inhabitants from returning to their lands and to protect the joint UK/US military base on Diego Garcia Island.

Other critiques include: their cost (higher than that of passive management), conflicts with human development goals, inadequate scope to address factors such as climate change and invasive species.

Recent Research

The larvae of the yellow tang can drift more than 100 miles and reseed in a distant location.

In 2010, one study found that fish larvae can drift on ocean currents and reseed fish stocks at a distant location. This finding demonstrated that fish populations can be connected to distant locations through the process of larval drift.

They investigated the yellow tang, because larva of this species stay in the general area of the reef in which they first settle. The tropical yellow tang is heavily fished by the aquarium trade. By the late 1990s, their stocks were collapsing. Nine MPAs were established off the coast of Hawaii to protect them. Larval drift has helped them establish themselves in different locations, and the fishery is recovering. "We've clearly shown that fish larvae that were spawned inside marine reserves can drift with currents and replenish fished areas long distances away," said coauthor Mark Hixon.

Marine Mammal Protection Act

The Marine Mammal Protection Act (MMPA) was the first act of the United States Congress to call specifically for a totem and taboo-style approach to wildlife management, setting aside marine mammals as untouchable and unapproachable. It is seen as the first federal animal rights legislation in the United States. It was signed into law on October 21, 1972 by President Richard Nixon and took effect 60 days later on December 21, 1972. MMPA prohibits the "taking" of marine mammals, and enacts a moratorium on the import, export, and sale of any marine mammal, along with any marine mammal part or product within the United States. The Act defines "take" as "the act of hunting, killing, capture, and/or harassment of any marine mammal; or, the attempt at such." The MMPA defines harassment as "any act of pursuit, torment or annoyance which has the potential to either: a. injure a marine mammal in the wild, or b. disturb a marine mammal by causing disruption of behavioral patterns, which includes, but is not limited to, migration, breathing, nursing, breeding, feeding, or sheltering." The MMPA provides for enforcement of its prohibitions, and for the issuance of regulations to implement its legislative goals.

One species of marine mammal: the Steller sea lion - this haul rests on rocks located on Amak Island.

Marine Mammal Management

Authority to manage the MMPA was divided between the Secretary of the Interior through the U.S. Fish and Wildlife Service (Service), and the Secretary of Commerce, which is delegated to the National Oceanic and Atmospheric Administration (NOAA). Subsequently, a third Federal agency, the Marine Mammal Commission (MMC), was established to review existing policies and make recommendations to the Service and the NOAA better implement the MMPA. Coordination between these three Federal agencies is necessary in order to provide the best management practices for marine mammals.

Under the MMPA, the Service is responsible for ensuring the protection of sea otters and marine otters, walruses, polar bears, three species of manatees, and dugongs. NOAA was given responsibility to conserve and manage pinnipeds including seals and sea lions and cetaceans such as whales and dolphins.

Marine Mammal Permits and International Coordination

The MMPA prohibits the take and exploitation of any marine mammal without appropriate authorization, which may only be given by the Service. Permits may be issued for scientific research, public display, and the importation/exportation of marine mammal parts and products upon determination by the Service that the issuance is consistent with the MMPA's regulations. The two types of permits issued by the National Marine Fisheries Service's Office of Protected Resources are incidental and directed. Incidental permits, which allow for some unintentional taking of small numbers of marine mammal, are granted to U.S. citizens who engage in a specified activity other than commercial fishing in a specified geographic area. Directed permits are required for any proposed marine mammal scientific research activity that involves taking marine mammals.

Applications for such permits are reviewed and issued the Service's Division of Management Authority, through the International Affairs office. This office also houses the Division of International Conservation, which is directly responsible for coordinating international activities for marine mammal species found in both U.S. and International waters, or are absent from U.S. waters. Marine mammal species inhabiting both U.S. and International waters include the West Indian manatee, sea otter, polar bear, and Pacific walrus. Species not present in U.S. waters include the West African and Amazonian manatee, dugong, Atlantic walrus, and marine otter.

Marine Mammal Conservation in the Field

In efforts to conserve and manage marine mammal species, the Service has appointed field staff dedicated to working with partners to conduct population censuses, assess population health, develop and implement conservation plans, promulgate regulations, and create cooperative relationships internationally.

Various Marine Mammal Management offices are located on either coast. The Service's Marine Mammal Management office in Anchorage, Alaska is responsible for the management and conservation of polar bears, Pacific walruses, and northern sea otters in Alaska. Northern sea otters present in Washington State are managed by the Western Washington Field Office, while southern sea otters residing in California are managed by the Ventura Field Office. West Indian manatee populations extend from Texas to Rhode Island, and are also present in the Caribbean Sea; however, this species is most prevalent near Florida (the Florida subspecies) and Puerto Rico (the Antillean subspecies). The Service's Jacksonville Field Office manages the Florida manatee, while the Boqueron Field Office manages the Antillean manatee.

The polar bear, southern sea otter, marine otter, all three species of manatees, and the dugong are also concurrently listed under the Endangered Species Act (ESA).

Amendments

Amendments enacted in 1981 established conditions for permits to be granted to take marine mammals "incidentally" in the course of commercial fishing. In addition, the amendments provided additional conditions and procedures for transferring management authority to the States, and authorized appropriations through FY 1984.

Policies Created in 1982

- Some marine mammal species or stocks may be in danger of extinction or depletion as a result of human activities
- These species or stocks must not be permitted to fall below their optimum sustainable population level (depleted)

- Measures should be taken to replenish these species or stocks

- There is inadequate knowledge of the ecology and population dynamics

- Marine mammals have proven to be resources of great international significance

The 1984 amendments established conditions to be satisfied as a basis for importing fish and fish products from nations engaged in harvesting yellowfin tuna with purse seines and other commercial fishing technology, as well as authorized appropriations for agency activities through FY 1988.

Amended in 1988

- the establishment of conditions and procedures for the Secretaries of Commerce and Interior to review the status of populations to determine if they should be listed as "depleted" (below optimal, sustainable population numbers or listed as threatened or endangered);

- the preparation of conservation plans for any species listed as depleted, including a requirement that such plans be modeled after recovery plans developed pursuant to the Endangered Species Act;

- the listing of conditions under which permits may be issued to take marine mammals for the protection and welfare of the animals, including importation, public display, scientific research, and enhancing the survival or recovery of a species; and

- a reward system under which the Secretary of the Treasury can pay up to $2500 to individuals providing information leading to convictions for violations of the Act.

Amended in 1995

- Certain exceptions to the take prohibitions, such as for Alaska Native subsistence and permits and authorizations for scientific research;

- A program to authorize and control the taking of marine mammals incidental to commercial fishing operations;

- Preparation of stock assessments for all marine mammal stocks in waters under U.S. jurisdiction; and

- Studies of pinniped-fishery interactions.

Findings

Congress found that: all species and population stocks of marine mammals are, or may be, in danger of extinction or depletion due to human activities; these mammals should not be permitted to diminish below their optimum sustainable population; measures should be taken immediately to replenish any of these mammals that have diminished below that level, and efforts should be made to protect essential habitats; there is inadequate knowledge of the ecology and population dynamics of these mammals; negotiations should be undertaken immediately to encourage international arrangements for research and conservation of these mammals. Congress declared that marine mammals are resources of great international significance (aesthetic, recreational and economic), and should be protected and encouraged to develop to the greatest extent feasible commensurate with sound policies of resource management. The primary management objective should be to maintain the health and stability of the marine ecosystem. The goal is to obtain an optimum sustainable population within the carrying capacity of the habitat.

Sustainable Fishery

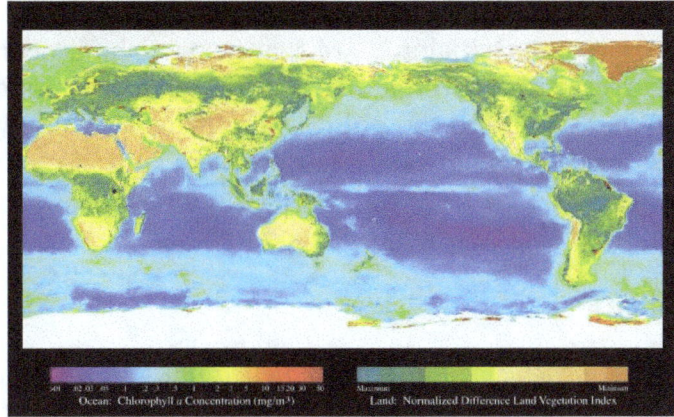

SeaWiFS map showing the levels of primary production in the world's oceans

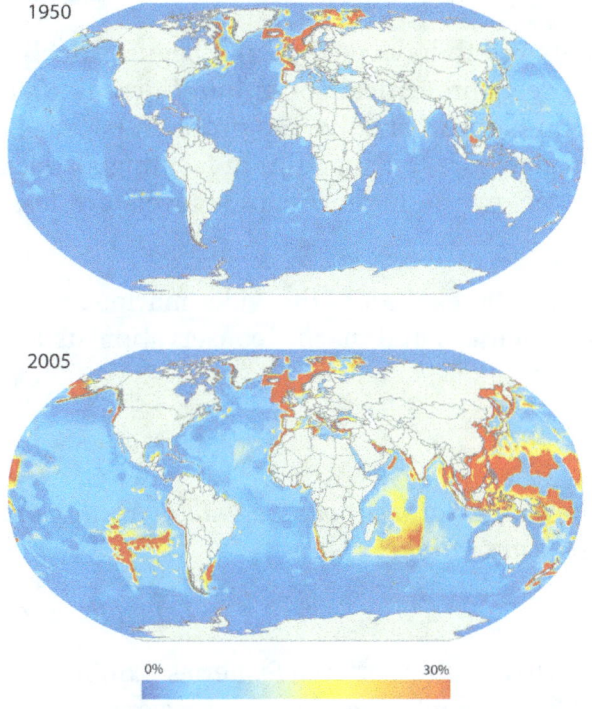

Primary production required (PPR) to sustain global marine fisheries landings expressed as percentage of local primary production (PP).Estimates of PPR, PP and PPR/PP computed per 0.5° latitude/longitude ocean cells. PPR estimates based on the catch database and PP estimates derived from SeaWiFS's global ocean colour satellite data. The maps represent total annual landings for 1950 (top) and 2005 (bottom). Note that PP estimates are static and derived from the synoptic observation for 1998.

A conventional idea of a sustainable fishery is that it is one that is harvested at a sustainable rate, where the fish population does not decline over time because of fishing practices. Sustainability in fisheries combines theoretical disciplines, such as the population dynamics of fisheries, with practical strategies, such as avoiding overfishing through techniques such as individual fishing quotas, curtailing destructive and illegal fishing practices by lobbying for appropriate law and policy, setting up protected areas, restoring collapsed fisheries, incorporating all externalities involved in harvesting marine ecosystems into fishery economics, educating stakeholders and the wider public, and developing independent certification programs.

Some primary concerns around sustainability are that heavy fishing pressures, such as overexploitation and growth or recruitment overfishing, will result in the loss of significant potential yield; that stock structure will erode to the point where it loses diversity and resilience to environmental fluctuations; that ecosystems and their economic infrastructures will cycle between collapse and recovery; with each cycle less productive than its predecessor; and that changes will occur in the trophic balance (fishing down marine food webs).

Overview

Sustainable management of fisheries cannot be achieved without an acceptance that the long-term goals of fisheries management are the same as those of environmental conservation

—Daniel Pauly and Dave Preikshot

Global wild fisheries are believed to have peaked and begun a decline, with valuable habitats, such as estuaries and coral reefs, in critical condition. Current aquaculture or farming of piscivorous fish, such as salmon, does not solve the problem because farmed piscivores are fed products from wild fish, such as forage fish. Salmon farming also has major negative impacts on wild salmon. Fish that occupy the higher trophic levels are less efficient sources of food energy.

Fishery ecosystems are an important subset of the wider marine environment. This section documents the views of fisheries scientists and marine conservationists about innovative approaches towards sustainable fisheries.

History

In the end, we will conserve only what we love; we will love only what we understand; and we will understand only what we are taught

—Senegalese conservationist Baba Dioum

In his 1883 inaugural address to the International Fisheries Exhibition in London, Thomas Huxley asserted that overfishing or "permanent exhaustion" was scientifically impossible, and stated that probably "all the great sea fisheries are inexhaustible". In reality, by 1883 marine fisheries were already collapsing. The United States Fish Commission was established 12 years earlier for the purpose of finding why fisheries in New England were declining. At the time of Huxley's address, the Atlantic halibut fishery had already collapsed (and has never recovered).

Traditional Management of Fisheries

Traditionally, fisheries management and the science underpinning it was distorted by its "narrow focus on target populations and the corresponding failure to account for ecosystem effects leading to declines of species abundance and diversity" and by perceiving the fishing industry as "the sole legitimate user, in effect the owner, of marine living resources." Historically, stock assessment scientists usually worked in government laboratories and considered their work to be providing services to the fishing industry. These scientists dismissed conservation issues and distanced themselves from the scientists and the science that raised the issues. This happened even as commercial fish stocks deteriorated, and even though many governments were signatories to binding conservation agreements.

Defining Sustainability

The notion of sustainable development is sometimes regarded as an unattainable, even illogical notion because development inevitably depletes and degrades the environment.

Ray Hilborn, of the University of Washington, distinguishes three ways of defining a sustainable fishery:

- *Long term constant yield* is the idea that undisturbed nature establishes a steady state that changes little over time. Properly done, fishing at up to maximum sustainable yield allows nature to adjust to a new steady state, without compromising future harvests. However, this view is naive, because constancy is not an attribute of marine ecosystems, which dooms this approach. Stock abundance fluctuates naturally, changing the potential yield over short and long term periods.

- *Preserving intergenerational equity* acknowledges natural fluctuations and regards as unsustainable only practices which damage the genetic structure destroy habitat, or deplete stock levels to the point where rebuilding requires more than a single generation. Providing rebuilding takes only one generation, overfishing may be economically foolish, but it is not unsustainable. This definition is widely accepted.

- *Maintaining a biological, social and economic system* considers the health of the human ecosystem as well as the marine ecosystem. A fishery which rotates among multiple species can deplete individual stocks and still be sustainable so long as the ecosystem retains its intrinsic integrity. Such a definition might consider as sustainable fishing practices that lead to the reduction and possible extinction of some species.

Social Sustainability

Fisheries and aquaculture are, directly or indirectly, a source of livelihood for over 500 million people, mostly in developing countries.

Social sustainability can conflict with biodiversity. A fishery is socially sustainable if the fishery ecosystem maintains the ability to deliver products the society can use. Major species shifts within the ecosystem could be acceptable as long as the flow of such products continues. Humans have been operating such regimes for thousands of years, transforming many ecosystems, depleting or driving to extinction many species.

According to Hilborn, the "loss of some species, and indeed transformation of the ecosystem is not incompatible with sustainable harvests." For example, in recent years, barndoor skates have been caught as bycatch in the western Atlantic. Their numbers have severely declined and they will probably go extinct if these catch rates continue. Even if the barndoor skate goes extinct, changing the ecosystem, there could still be sustainable fishing of other commercial species.

> " *To a great extent, sustainability is like good art, it is hard to describe but we know it when we see it.* "
>
> — *Ray Hilborn*

Reconciling Fisheries with Conservation

Management goals might consider the impact of salmon on bear and river ecosystems.

At the Fourth World Fisheries Congress in 2004, Daniel Pauly asked, "How can fisheries science and conservation biology achieve a reconciliation?", then answered his own question, "By accepting each other's essentials: that fishing should remain a viable occupation; and that aquatic ecosystems and their biodiversity are allowed to persist."

A relatively new concept is relationship farming. This is a way of operating farms so they restore the food chain in their area. Re-establishing a healthy food chain can result in the farm automatically filtering out impurities from feed water and air, feeding its own food chain, and additionally producing high net yields for harvesting. An example is the large cattle ranch Veta La Palma in southern Spain. Relationship farming was first made popular by Joel Salatin who created a 220 hectare relationship farm featured prominently in Michael Pollan's book *The Omnivore's Dilemma* (2006) and the documentary films, Food, Inc. and Fresh. The basic concept of relationship farming is to put effort into building a healthy food chain, and then the food chain does the hard work.

Obstacles

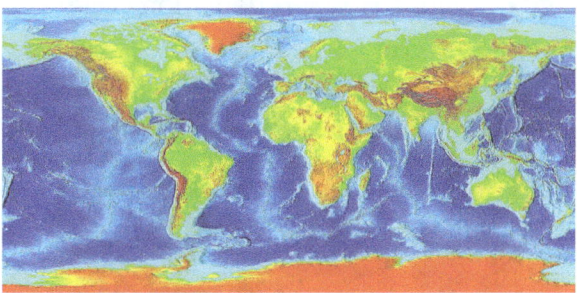

Large areas of the global continental shelf, highlighted in cyan, have had heavy bottom trawls repeatedly dragged over them.

Overfishing

Overfishing can be sustainable. According to Hilborn, overfishing can be "a misallocation of societies' resources", but it does not necessarily threaten conservation or sustainability".

Fishing down the food web

Overfishing is traditionally defined as harvesting so many fish that the yield is less than it would be if fishing were reduced. For example, Pacific salmon are usually managed by trying to determine how many spawning salmon, called the "escapement", are needed each generation to produce the maximum harvestable surplus. The optimum escapement is that needed to reach that surplus. If the escapement is half the optimum, then normal fishing looks like overfishing. But this is still sustainable fishing, which could continue indefinitely at its reduced stock numbers and yield. There is a wide range of escapement sizes that present no threat that the stock might collapse or that the stock structure might erode.

On the other hand, overfishing can precede severe stock depletion and fishery collapse. Hilborn points out that continuing to exert fishing pressure while production decreases, stock collapses and the fishery fails, is largely "the product of institutional failure."

Today over 70% of fish species are either fully exploited, overexploited, depleted, or recovering from depletion. If overfishing does not decrease, it is predicted that stocks of all species currently commercially fished for will collapse by 2048."

A Hubbert linearization (Hubbert curve) has been applied to the whaling industry, as well as charting the price of caviar, which depends on sturgeon stocks. Another example is North Sea cod. Comparing fisheries and mineral extraction tells us that human pressure on the environment is causing a wide range of resources to go through a Hubbert depletion cycle.

Coastal fishing communities in Bangladesh are vulnerable to flooding from sea-level rises.

Island with fringing reef in the Maldives. Coral reefs are dying around the world.

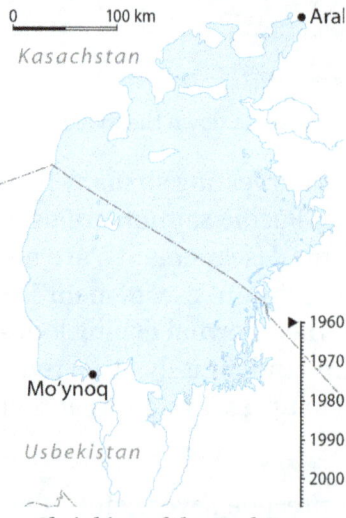

Shrinking of the Aral Sea

Habitat Modification

Nearly all the world's continental shelves, and large areas of continental slopes, underwater ridges, and seamounts, have had heavy bottom trawls and dredges repeatedly dragged over their surfaces. For fifty years, governments and organizations, such as the Asian Development Bank, have encouraged the fishing industry to develop trawler fleets. Repeated bottom trawling and dredging literally flattens diversity in the benthic habitat, radically changing the associated communities.

Changing the Ecosystem Balance

Since 1950, 90 percent of 25 species of big predator fish have gone.

- How we are emptying our seas, *The Sunday Times*, May 10, 2009.

- Pauly, Daniel (2004) Reconciling Fisheries with Conservation: the Challenge of Managing Aquatic Ecosystems Fourth World Fisheries Congress, Vancouver, 2004.

Climate Change

Rising ocean temperatures and ocean acidification are radically altering aquatic ecosystems. Climate change is modifying fish distribution and the productivity of marine and freshwater species. This reduces sustainable catch levels across many habitats, puts pressure on resources needed for aquaculture, on the communities that depend on fisheries, and on the oceans' ability to capture and store carbon (biological pump). Sea level rise puts coastal fishing communities at risk, while changing rainfall patterns and water use impact on inland (freshwater) fisheries and aquaculture.

Ocean Pollution

A recent survey of global ocean health concluded that all parts of the ocean have been impacted by human development and that 41 percent has been fouled with human polluted runoff, overfishing, and other abuses. Pollution is not easy to fix, because pollution sources are so dispersed, and are built into the economic systems we depend on.

The United Nations Environment Programme (UNEP) mapped the impacts of stressors such as climate change, pollution, exotic species, and over-exploitation of resources on the oceans. The report shows at least 75 percent of the world's key fishing grounds may be affected.

Diseases and Toxins

Large predator fish contain significant amounts of mercury, a neurotoxin which can affect fetal development, memory, mental focus, and produce tremors.

Irrigation

Abandoned ship near Aral, Kazakhstan

Lakes are dependent on the inflow of water from its drainage basin. In some areas, aggressive irrigation has caused this inflow to decrease significantly, causing water depletion and a shrinking of the lake. The most notable example is the Aral Sea, formerly among the four largest lakes in the world, now only a tenth of its former surface area.

Remediation

Fisheries Management

Fisheries management draws on fisheries science to enable sustainable exploitation. Modern fisheries management is often defined as mandatory rules based on concrete objectives and a mix of management techniques, enforced by a monitoring control and surveillance system.

- Ideas and rules: Economist Paul Romer believes sustainable growth is possible providing the right ideas (technology) are combined with the right rules, rather than simply hectoring fishers. There has been no lack of innovative ideas about how to harvest fish. He characterizes failures as primarily failures to apply appropriate rules.

- Fishing subsidies: Government subsidies influence many of the world fisheries. Operating cost subsidies allow European and Asian fishing fleets to fish in distant waters, such as West Africa. Many experts reject fishing subsidies and advocate restructuring incentives globally to help struggling fisheries recover.

- Economics: Another focus of conservationists is on curtailing detrimental human activities by improving fisheries' market structure with techniques such as salable fishing quotas, like those set up by the Northwest Atlantic Fisheries Organization, or laws such as those listed below.

- Payment for Ecosystem Services: Environmental Economist, Essam Y Mohammed, argues that by creating direct economic incentives, whereby people are able to receive payment for the services their property provides, will help to establish sustainable fisheries around the world as well as inspire conservation where it otherwise would not.

- Sustainable fisheries certification: A promising direction is the independent certification programs for sustainable fisheries conducted by organizations such as the Marine Stewardship Council and Friend of the Sea. These programs work at raising consumer awareness and insight into the nature of their seafood purchases.

- Ecosystem based fisheries.

Ecosystem Based Fisheries

We propose that rebuilding ecosystems, and not sustainability per se, should be the goal of fishery management. Sustainability is a deceptive goal because human harvesting of fish leads to a progressive simplification of ecosystems in favour of smaller, high turnover, lower trophic level fish species that are adapted to withstand disturbance and habitat degradation.

—Tony Pitcher and Daniel Pauly

According to marine ecologist Chris Frid, the fishing industry points to marine pollution and global warming as the causes of recent, unprecedented declines in fish populations. Frid counters that overfishing has also altered the way the ecosystem works. "Everybody would like to see the rebuilding of fish stocks and this can only be achieved if we understand all of the influences, human and natural, on fish dynamics." He adds: "fish communities can be altered in a number of ways, for example they can decrease if particular-sized individuals of a species are targeted, as this affects predator and prey dynamics. Fishing, however, is not the sole cause of changes to marine life—pollution is another example....No one factor operates in isolation and components of the ecosystem respond differently to each individual factor."

The traditional approach to fisheries science and management has been to focus on a single species. This can be contrasted with the ecosystem-based approach. Ecosystem-based fishery concepts have been implemented in some regions. In a 2007 effort to "stimulate much needed discussion" and "clarify the essential components" of ecosystem-based fisheries science, a group of scientists offered the following ten commandments for ecosystem-based fisheries scientists

- Keep a perspective that is holistic, risk-adverse and adaptive.

- Maintain an "old growth" structure in fish populations, since big, old and fat female fish have been shown to be the best spawners, but are also susceptible to overfishing.

- Characterize and maintain the natural spatial structure of fish stocks, so that management boundaries match natural boundaries in the sea.

- Monitor and maintain seafloor habitats to make sure fish have food and shelter.

- Maintain resilient ecosystems that are able to withstand occasional shocks.

- Identify and maintain critical food-web connections, including predators and forage species.

- Adapt to ecosystem changes through time, both short-term and on longer cycles of decades or centuries, including global climate change.

- Account for evolutionary changes caused by fishing, which tends to remove large, older fish.

- Include the actions of humans and their social and economic systems in all ecological equations.

Marine Protected Areas

Strategies and techniques for marine conservation tend to combine theoretical disciplines, such as population biology, with practical conservation strategies, such as setting up protected areas, as with Marine Protected Areas (MPAs) or Voluntary Marine Conservation Areas. Each nation defines MPAs independently, but they commonly involve increased protection for the area from fishing and other threats.

Marine life is not evenly distributed in the oceans. Most of the really valuable ecosystems are in relatively shallow coastal waters, above or near the continental shelf, where the sunlit waters are often nutrient rich from land runoff or upwellings at the continental edge, allowing photosynthesis, which energizes the lowest trophic levels. In the 1970s, for reasons more to do with oil drilling than with fishing, the U.S. extended its jurisdiction, then 12 miles from the coast, to 200 miles. This made huge shelf areas part of its territory. Other nations followed, extending national control to what became known as the exclusive economic zone (EEZ). This move has had many implications for fisheries conservation, since it means that most of the most productive maritime ecosystems are now under national jurisdictions, opening possibilities for protecting these ecosystems by passing appropriate laws.

Daniel Pauly characterises marine protected areas as "a conservation tool of revolutionary importance that is being incorporated into the fisheries mainstream." The Pew Charitable Trusts have funded various initiatives aimed at encouraging the development of MPAs and other ocean conservation measures.

Fish Farming

There exists concerns that farmed fish cannot produce necessary yields efficiently. For example, farmed salmon eat three pounds of wild fish to produce one pound of salmon.

Laws and Treaties

International laws and treaties related to marine conservation include the 1966 Convention on Fishing and Conservation of Living Resources of the High Seas. United States laws related to marine conservation include the 1972 Marine Mammal Protection Act, as well as the 1972 Marine Protection, Research and Sanctuaries Act which established the National Marine Sanctuaries program. Magnuson-Stevens Fishery Conservation and Management Act.

Awareness Campaigns

Various organizations promote sustainable fishing strategies, educate the public and stakeholders, and lobby for conservation law and policy. The list includes the Marine Conservation Biology Institute and Blue Frontier Campaign in the U.S., The U.K.'s Frontier (the Society for Environmental Exploration) and Marine Conservation Society, Australian Marine Conservation Society, International Council for the Exploration of the Sea (ICES), Langkawi Declaration, Oceana, PROFISH, and the Sea Around Us Project, International Collective in Support of Fishworkers, World Forum of Fish Harvesters and Fish Workers, Frozen at Sea Fillets Association and CEDO.

Introducing the results of long term monitoring to a local fishermen in Kihnu, Estonia.

The United Nations Millennium Development Goals include, as goal #7: target 2, the intention to "reduce biodiversity loss, achieving, by 2010, a significant reduction in the rate of loss", including improving fisheries management to reduce depletion of fish stocks.

Some organizations certify fishing industry players for sustainable or good practices, such as the Marine Stewardship Council and Friend of the Sea.

Other organizations offer advice to members of the public who eat with an eye to sustainability. According to the marine conservation biologist Callum Roberts, four criteria apply when choosing seafood:

- Is the species in trouble in the wild where the animals were caught?

- Does fishing for the species damage ocean habitats?

- Is there a large amount of bycatch taken with the target species?

- Does the fishery have a problem with discards—generally, undersized animals caught and thrown away because their market value is low?

The following organizations have download links for wallet-sized cards, listing good and bad choices:

- Monterey Bay Aquarium Seafood Watch, USA

- Blue Ocean Institute, USA

- Marine Conservation Society, UK

- Australian Marine Conservation Society

- The Southern African Sustainable Seafood Initiative

Data Issues

Data Quality

One of the major impediments to the rational control of marine resources is inadequate data. According to fisheries scientist Milo Adkison (2007), the primary limitation in fisheries management decisions is poor data. Fisheries management decisions are often based on population models, but the models need quality data to be accurate. Scientists and fishery managers would be better served with simpler models and improved data.

Unreported Fishing

Estimates of illegal catch losses range between $10 billion and $23 billion annually, representing between 11 and 26 million tonnes.

- Incidental catch

Shifting Baselines

Shifting baselines is a term which describes the way significant changes to a system are measured against previous baselines, which themselves may represent significant changes from the original state of the system. The term was first used by the fisheries scientist Daniel Pauly in his paper "Anecdotes and the shifting baseline syndrome of fisheries". Pauly developed the term in reference to fisheries management where fisheries scientists sometimes fail to identify the correct "baseline" population size (e.g. how abundant a fish species population was before human exploitation) and thus work with a shifted baseline. He describes the way that radically depleted fisheries were evaluated by experts who used the state of the fishery at the start of their careers as the baseline, rather than the fishery in its untouched state. Areas that swarmed with a particular species hundreds of years ago, may have experienced long term decline, but it is the level of decades previously that is considered the appropriate reference point for current populations. In this way large declines in ecosystems or species over long periods of time were, and are, masked. There is a loss of perception of change that occurs when each generation redefines what is "natural".

Looting the Seas

Looting the seas is the name given by the International Consortium of Investigative Journalists to a series of journalistic investigations into areas directly affecting the sustainability of fisheries. So far they have investigated three areas involving fraud, negligence and overfishing:

- The black market in bluefin tuna

- Subsidies propping up the Spanish fishing industry

- Overfishing of the southern jack mackerel

Other Factors

The focus of sustainable fishing is often on the fish. Other factors are sometimes included in the broader question of sustainability. The use of non-renewable resources is not fully sustainable. This might include diesel fuel for the fishing ships and boats: there is even a debate about the long term sustainability of biofuels. Modern fishing nets are usually made of artificial polyamides like nylon. Synthetic braided ropes are generally made from nylon, polyester, polypropylene or high performance fibers such as high modulus polyethylene (HMPE) and aramid.

Energy and resources are employed in fish processing, refrigeration, packaging, logistics, etc. The methodologies of Life-cycle assessment are useful to evaluate the sustainability of components and systems. These are part of the broad question of sustainability.

References

- Pauly, D; Christensen, V; Guénette, S; Pitcher, TJ; Sumaila, UR; Walters, CJ; Watson, R; Zeller, D (2002). "Towards sustainability in world fisheries". Nature. 418: 689–695. doi:10.1038/nature01017

- Ray, G. Carleton (2004) "Issues and Mechanisms", part 1 in Coastal-marine Conservation: Science and Policy. Malden, MA: Blackwell Pub. ISBN 978-0-632-05537-1

- Thys, Tierney (30 November 2003). "Tracking Ocean Sunfish, Mola mola with Pop-Up Satellite Archival Tags in California Waters". OceanSunfish.org. Retrieved 14 June 2007

- Jenkins, Lekeliad (2010). "Profile and influence of the successful fisher-Inventor of marine conservation technology". Conservation and Society. 8: 44. doi:10.4103/0972-4923.62677

- Carleton, Ray G.; McCormick, Jerry (1 April 2009). Coastal-Marine Conservation: Science and Policy. John Wiley & Sons. ISBN 978-1-4443-1124-2

- Berezansky, L.; Idels, L.; Kipnis, M. (2011). "Mathematical model of marine protected areas". IMA Journal of Applied Mathematics. 76 (2): 312–325. ISSN 0272-4960. doi:10.1093/imamat/hxq043

- Hughes, TP; Bellwooda, DR; Folkeb, C; Steneck, RS; Wilson, J (2005). "New paradigms for supporting the resilience of marine ecosystems". Trends in Ecology & Evolution. 20 (7): 380–386. doi:10.1016/j.tree.2005.03.022

- Marine Protected Areas (MPA) (2010-10-15). "Areas of Biodiversity Importance: Marine Protected Areas, 2010". Biodiversitya-z.org. Retrieved 2012-06-07

- Halpern, B. (2003). "The impact of marine reserves: do reserves work and does reserve size matter?". Ecological Applications. 13: S117– S137. doi:10.1890/1051-0761(2003)013[0117:TIOMRD]2.0.CO;2

- Berkes F, Mahon R, McConney P, Pollnac R and Pomeroy R (2001) Managing Small-Scale Fisheries: Alternative Directions and Methods IDRC,. ISBN 978-0-88936-943-6

- "Marine Protected Areas Government Website". Mpa.gov. 2012-05-07. Archived from the original on 2012-06-06. Retrieved 2012-06-07

- Pandolfi, J.M.; et al. (2003). "Global trajectories of long-term decline of coral reef ecosystems". Science. 301 (5635): 955–958. PMID 12920296. doi:10.1126/science.1085706

- "Conservation International – World's Largest Marine Protected Area Created in Pacific Ocean". Conservation. org. Retrieved 2012-06-07

Permissions

Index

www.ingramcontent.com/pod-product-compliance
Lightning Source LLC
Chambersburg PA
CBHW080408190526
45161CB00003B/166